IFIP Advances in Information
and Communication Technology

306

IFIP – The International Federation for Information Processing

IFIP was founded in 1960 under the auspices of UNESCO, following the First World Computer Congress held in Paris the previous year. An umbrella organization for societies working in information processing, IFIP's aim is two-fold: to support information processing within its member countries and to encourage technology transfer to developing nations. As its mission statement clearly states,

> *IFIP's mission is to be the leading, truly international, apolitical organization which encourages and assists in the development, exploitation and application of information technology for the benefit of all people.*

IFIP is a non-profitmaking organization, run almost solely by 2500 volunteers. It operates through a number of technical committees, which organize events and publications. IFIP's events range from an international congress to local seminars, but the most important are:

- The IFIP World Computer Congress, held every second year;
- Open conferences;
- Working conferences.

The flagship event is the IFIP World Computer Congress, at which both invited and contributed papers are presented. Contributed papers are rigorously refereed and the rejection rate is high.

As with the Congress, participation in the open conferences is open to all and papers may be invited or submitted. Again, submitted papers are stringently refereed.

The working conferences are structured differently. They are usually run by a working group and attendance is small and by invitation only. Their purpose is to create an atmosphere conducive to innovation and development. Refereeing is less rigorous and papers are subjected to extensive group discussion.

Publications arising from IFIP events vary. The papers presented at the IFIP World Computer Congress and at open conferences are published as conference proceedings, while the results of the working conferences are often published as collections of selected and edited papers.

Any national society whose primary activity is in information may apply to become a full member of IFIP, although full membership is restricted to one society per country. Full members are entitled to vote at the annual General Assembly, National societies preferring a less committed involvement may apply for associate or corresponding membership. Associate members enjoy the same benefits as full members, but without voting rights. Corresponding members are not represented in IFIP bodies. Affiliated membership is open to non-national societies, and individual and honorary membership schemes are also offered.

Gilbert Peterson   Sujeet Shenoi (Eds.)

Advances in Digital Forensics V

Fifth IFIP WG 11.9 International Conference
on Digital Forensics
Orlando, Florida, USA, January 26-28, 2009
Revised Selected Papers

 Springer

Volume Editors

Gilbert Peterson
Air Force Institute of Technology
Wright-Patterson Air Force Base, OH 45433-7765, USA
E-mail: gilbert.peterson@afit.edu

Sujeet Shenoi
University of Tulsa
Tulsa, OK 74104-3189, USA
E-mail: sujeet@utulsa.edu

CR Subject Classification (1998): H.2.8, I.5, C.2, I.5.2, I.7.5, G.3

ISSN   1868-4238

ISBN-13   978-3-642-26018-6 Springer Berlin Heidelberg New York

springer.com

© IFIP International Federation for Information Processing 2009
Softcover reprint of the hardcover 1st edition 2009

Typesetting: Camera-ready by author, data conversion by Scientific Publishing Services, Chennai, India
Printed on acid-free paper   SPIN: 12753710   06/3180   5 4 3 2 1 0

Contents

PART VI INVESTIGATIVE TECHNIQUES

PART VII LEGAL ISSUES

PART VIII EVIDENCE MANAGEMENT

Contributing Authors

Yuki Ashino received his Ph.D. degree in Computer Engineering from Tokyo Denki University, Tokyo, Japan. He is currently a researcher at NEC, Tokyo, Japan. His research interests are in the area of digital forensics.

Jason Beckett is a Ph.D. student in Forensic Computing at the Defence and Systems Institute Safeguarding Australia Research Laboratory at the University of South Australia, Adelaide, Australia. His research interests include forensic computing validation and verification.

Nicole Beebe is an Assistant Professor of Information Systems and Technology Management at the University of Texas at San Antonio, San Antonio, Texas. Her research interests include digital forensics, information security and data mining.

David Billard is a Professor of Computer Science at the University of Applied Sciences of Western Switzerland, Geneva, Switzerland. His research interests include e-discovery, cell phone and small-scale device forensics, and forensic models and procedures.

Thomas Breuel is a Professor of Computer Science at the Technical University of Kaiserslautern, Kaiserslautern, Germany. He also leads the Image Understanding and Pattern Recognition Research Group at the German Research Center for Artificial Intelligence, Kaiserslautern, Germany. His research interests include large-scale, real-world pattern recognition and machine learning, multimedia content analysis and document, image and video forensics.

Jonathan Butts is a Ph.D. student in Computer Science at the University of Tulsa, Tulsa, Oklahoma. His research interests include information assurance, critical infrastructure protection and embedded devices.

Kam-Pui Chow is an Associate Professor of Computer Science at the University of Hong Kong, Hong Kong, China. His research interests include information security, digital forensics, live system forensics and digital surveillance.

Fred Cohen is the Chief Executive Officer of Fred Cohen and Associates, and the President of California Sciences Institute, Livermore, California. His research interests include information assurance, critical infrastructure protection and digital forensics.

Scott Conrad is a Research Assistant at the National Center for Forensic Science, University of Central Florida, Orlando, Florida. His research interests include personal gaming/entertainment devices and virtualization technologies.

Philip Craiger is an Assistant Professor of Engineering Technology and Assistant Director for Digital Evidence at the National Center for Forensic Science, University of Central Florida, Orlando, Florida. His research interests include information assurance and digital forensics.

David Dampier is an Associate Professor of Computer Science and Engineering, and the Director of the Southeast Region Forensics Training Center at Mississippi State University, Mississippi State, Mississippi. His research interests include digital forensics, information assurance and software engineering.

Greg Dorn is a Research Assistant at the National Center for Forensic Science, University of Central Florida, Orlando, Florida. His research interests include virtualization technologies and personal gaming/entertainment devices.

Keisuke Fujita is an M.S. student in Computer Engineering at Tokyo Denki University, Tokyo, Japan. His research interests are in the area of digital forensics.

Maiko Furusawa is an M.S. student in Computer Engineering at Tokyo Denki University, Tokyo, Japan. Her research interests are in the area of digital forensics.

Ashish Gehani is a Computer Scientist at SRI's Computer Science Laboratory in Menlo Park, California. His research interests include data lineage, digital forensics, risk management and intrusion response.

Pavel Gladyshev is a Lecturer of Computer Science and Informatics at University College Dublin, Dublin, Ireland. His research interests include information security and digital forensics.

Yong Guan is an Associate Professor of Electrical and Computer Engineering at Iowa State University, Ames, Iowa. His research interests include Internet security and privacy.

Marianne Hoebich received her M.S. degree in Information Assurance and Security from Purdue University, West Lafayette, Indiana. Her research interests are in the areas of information assurance and digital forensics.

Ricci Ieong is a Ph.D. student in Computer Science at the University of Hong Kong, Hong Kong, China. His research interests include digital forensics, peer-to-peer forensics and time correlation analysis.

Ibrahim Imam is an Associate Professor of Computer Engineering and Computer Science at the University of Louisville, Louisville, Kentucky. His research interests include digital forensics and mathematics.

Joshua James is an M.Sc. student at the Center for Cybercrime Investigation, University College Dublin, Dublin, Ireland. His research interests include cybercrime investigation process models and standards, evidence correlation techniques and event reconstruction.

Florent Kirchner was an International Postdoctoral Fellow at SRI's Computer Science Laboratory in Menlo Park, California. His research interests include programming language semantics, formal verification of aeronautics systems and distributed proof frameworks.

Hugo Kleinhans is a Ph.D. student in Computer Science at the University of Tulsa, Tulsa, Oklahoma. His research interests include distributed systems, critical infrastructure protection, digital forensics and cyber policy.

Xiangwei Kong is a Professor of Electronic and Information Engineering at Dalian University of Technology, Dalian, China. Her research interests include multimedia security, image forensics, image processing and wireless sensor networks.

Michael Kwan is a Ph.D. student in Computer Science at the University of Hong Kong, Hong Kong, China. His research interests include digital forensics, digital evidence evaluation and the application of probabilistic models in digital forensics.

Pierre Lai is a Ph.D. student in Computer Science at the University of Hong Kong, Hong Kong, China. Her research interests include cryptography, peer-to-peer networks and digital forensics.

Frank Law is a Ph.D. student in Computer Science at the University of Hong Kong, Hong Kong, China. His research interests include digital forensics and time analysis.

Sydney Liles is a Ph.D. student in Computer Forensics at Purdue University, West Lafayette, Indiana. Her research interests include digital forensics and public policy.

Paul Lin is a Professor of Forensic Computing at the Central Police University, Taoyuan, Taiwan. His research interests include information assurance and digital forensics.

Yi-Chi Lin is an Electronic Evidence Support Officer at the South Australian Police E-Crime Laboratory, Adelaide, Australia. His research interests include forensic computing and culture.

Michael Losavio is an Instructor of Computer Engineering and Computer Science, and Justice Administration at the University of Louisville, Louisville, Kentucky. His research interests include legal and social issues in computing and digital crime.

Chris Marberry is a Senior Digital Evidence Research Assistant at the National Center for Forensic Science, University of Central Florida, Orlando, Florida. His research interests include digital forensics, computer security and virtualization technologies.

Jeff Marean is a retired Police Officer who resides in Louisville, Kentucky. His interests include network administration, information security and digital forensics.

Lodovico Marziale is a Ph.D. student in Computer Science at the University of New Orleans, New Orleans, Louisiana. His research interests include digital forensics, machine learning, and parallel and concurrent programming.

Nick Miles is an M.S. student in Computer Engineering and Computer Science at the University of Louisville, Louisville, Kentucky. His research interests are in the area of network security.

Robert Mills is an Assistant Professor of Electrical Engineering at the Air Force Institute of Technology, Wright-Patterson Air Force Base, Ohio. His research interests include network management and security, cyber warfare, communication systems and systems engineering.

Barry Mullins is an Assistant Professor of Computer Engineering at the Air Force Institute of Technology, Wright-Patterson Air Force Base, Ohio. His research interests include cyber operations, computer and network security, computer communication networks, embedded (sensor) and wireless networking, and reconfigurable computing systems.

Olfa Nasraoui is an Associate Professor of Computer Engineering and Computer Science at the University of Louisville, Louisville, Kentucky. Her research interests include data mining, web mining, stream data mining, personalization and pattern recognition

Richard Overill is a Senior Lecturer in Computer Science at King's College London, London, United Kingdom. His research interests include digital forensics, cybercrime analysis, anomaly detection, cyberwarfare and information security management.

Yanlin Peng is a Ph.D. candidate in Electrical and Computer Engineering at Iowa State University, Ames, Iowa. Her research interests include web security, data mining and distributed computing.

Gilbert Peterson is an Associate Professor of Computer Engineering at the Air Force Institute of Technology, Wright-Patterson Air Force Base, Ohio. His research interests include digital forensics, steganography, robotics and machine learning.

Mark Pollitt, Chair, IFIP Working Group 11.9 on Digital Forensics, is a faculty member in Engineering Technology and a principal with the National Center for Forensic Science, University of Central Florida, Orlando, Florida. His research interests include forensic processes, knowledge management, information security and forensic quality management.

Kamil Reddy is a Ph.D. student in Computer Science at the University of Pretoria, Pretoria, South Africa. His research interests include information privacy, digital forensics and information security.

Golden Richard is a Professor of Computer Science at the University of New Orleans, New Orleans, Louisiana, and the co-founder of Digital Forensics Solutions, LLC, New Orleans, Louisiana. His research interests include digital forensics, mobile computing and operating systems internals.

Carlos Rodriguez was a Research Assistant at the National Center for Forensic Science, University of Central Florida, Orlando, Florida. His research interests include personal gaming/entertainment devices.

Marcus Rogers is a Professor of Computer and Information Technology at Purdue University, West Lafayette, Indiana. His research interests include psychological digital crime scene analysis, applied behavioral profiling and digital evidence process models.

Vassil Roussev is an Assistant Professor of Computer Science at the University of New Orleans, New Orleans, Louisiana. His research interests include digital forensics, high-performance computing, distributed collaboration and software engineering.

Ryoichi Sasaki is a Professor of Information Security at Tokyo Denki University, Tokyo, Japan. His research interests include risk assessment, cryptography and digital forensics.

Karl Schrader is an M.S. student in Computer Engineering at the Air Force Institute of Technology, Wright-Patterson Air Force Base, Ohio. His research interests include digital forensics, peer-to-peer networking and digital cryptography.

Marco Schreyer is an M.Sc. student at the German Research Center for Artificial Intelligence, Kaiserslautern, Germany. His research interests include document forensics and investigative data mining.

Christian Schulze is a Researcher at the German Research Center for Artificial Intelligence, Kaiserslautern, Germany. His research interests include image processing, image/video retrieval, neural networks and genetic algorithms.

Natarajan Shankar is a Staff Scientist at SRI's Computer Science Laboratory in Menlo Park, California. His research interests include formal methods for hardware and software specification and verification, and automated deduction.

Sujeet Shenoi is the F.P. Walter Professor of Computer Science at the University of Tulsa, Tulsa, Oklahoma. His research interests include information assurance, digital forensics, critical infrastructure protection and intelligent control.

Jill Slay is the Director of the Defence and Systems Institute Safeguarding Australia Research Laboratory at the University of South Australia, Adelaide, Australia. Her research interests include information assurance, digital forensics, critical infrastructure protection and complex system modeling.

Armin Stahl is a Senior Researcher at the German Research Center for Artificial Intelligence, Kaiserslautern, Germany. His research interests include case-based reasoning, learning systems, electronic commerce, information retrieval and knowledge management.

April Tanner is a Ph.D. candidate in Computer Science at Mississippi State University, Mississippi State, Mississippi. Her research interests include digital forensics and information assurance.

Vincent Thacker is a Software Engineer with Alliant Technologies in Louisville, Kentucky. His research interests include peer-to-peer networks, networking and computer graphics.

Hayson Tse is a Ph.D. student in Computer Science at the University of Hong Kong, Hong Kong, China. His research interests are in the area of digital forensics.

Kenneth Tse is an M.Phil. student in Computer Science at the University of Hong Kong, Hong Kong, China. His research interests include digital forensics, file systems and data visualization.

Benjamin Turnbull is a Postdoctoral Research Fellow at the Defence and Systems Institute Safeguarding Australia Research Laboratory at the University of South Australia, Adelaide, Australia. His research interests include forensic computing, wireless technologies and drug-crime-related electronic analysis.

Tetsutaro Uehara is an Associate Professor of Computing and Media Studies at Kyoto University, Kyoto, Japan. His research interests include high-performance computing, secure multimedia streaming technology, network security and digital forensics.

Hein Venter is an Associate Professor of Computer Science at the University of Pretoria, Pretoria, South Africa. His research interests include network security, digital forensics, information privacy and intrusion detection.

Bo Wang is a Ph.D. student in Electronic and Information Engineering at Dalian University of Technology, Dalian, China. His research interests include multimedia security, image forensics and image processing.

Liqiang Wang received his M.S. degree in Computer Science from the University of New Orleans, New Orleans, Louisiana. His research interests are in the area of high-performance computing.

Roman Yampolskiy is an Assistant Professor of Computer Engineering and Computer Science at the University of Louisville, Louisville, Kentucky. His research interests include digital forensics, behavioral biometrics, pattern recognition, genetic algorithms, neural networks, artificial intelligence and games.

Xingang You is a Professor of Electrical Engineering at Beijing Electrical Technology Applications Institute, Beijing, China. His research interests include multimedia security, signal processing and communications.

Linfeng Zhang received his Ph.D. degree in Computer Engineering from Iowa State University, Ames, Iowa. His research interests include digital forensics and data streaming algorithms.

Yuandong Zhu is a Ph.D. student in Computer Science and Informatics at University College Dublin, Dublin, Ireland. His research interests include user activity analysis and forensic tool development.

Preface

Digital forensics deals with the acquisition, preservation, examination, analysis and presentation of electronic evidence. Networked computing, wireless communications and portable electronic devices have expanded the role of digital forensics beyond traditional computer crime investigations. Practically every type of crime now involves some aspect of digital evidence; digital forensics provides the techniques and tools to articulate this evidence in legal proceedings. Digital forensics also has myriad intelligence applications; furthermore, it has a vital role in information assurance – investigations of security breaches yield valuable information that can be used to design more secure and resilient systems.

This book, *Advances in Digital Forensics V*, is the fifth volume in the annual series produced by IFIP Working Group 11.9 on Digital Forensics, an international community of scientists, engineers and practitioners dedicated to advancing the state of the art of research and practice in digital forensics. The book presents original research results and innovative applications in digital forensics. Also, it highlights some of the major technical and legal issues related to digital evidence and electronic crime investigations.

This volume contains twenty-three edited papers from the Fifth IFIP WG 11.9 International Conference on Digital Forensics, held at the National Center for Forensic Science, Orlando, Florida, January 26–28, 2009. The papers were refereed by members of IFIP Working Group 11.9 and other internationally-recognized experts in digital forensics.

The chapters are organized into eight sections: themes and issues, forensic techniques, integrity and privacy, network forensics, forensic computing, investigative techniques, legal issues and evidence management. The coverage of topics highlights the richness and vitality of the discipline, and offers promising avenues for future research in digital forensics.

This book is the result of the combined efforts of several individuals. In particular, we thank Jonathan Butts, Rodrigo Chandia and Anita Presley for their tireless work on behalf of IFIP Working Group 11.9.

We also acknowledge the support provided by the National Center for Forensic Science, National Security Agency, Immigration and Customs Enforcement, and U.S. Secret Service.

GILBERT PETERSON AND SUJEET SHENOI

I

THEMES AND ISSUES

Chapter 1

DIGITAL FORENSICS AS A SURREAL NARRATIVE

Mark Pollitt

Abstract   Digital forensics is traditionally approached either as a computer science problem or as an investigative problem. In both cases, the goal is usually the same: attempt to locate discrete pieces of information that are probative. In the computer science approach, characteristics of the data are utilized to include or exclude objects, data or metadata. The investigative approach reviews the content of the evidence to interpret the data in the light of known facts and elements of the crime in order to determine probative information or information of lead value. This paper explores two literary theories, narrative theory and surrealism, for potential application to the digital forensic process. Narrative theory focuses on the "story" that is represented by text. At some level, a storage device may be viewed as a series of interweaving, possibly multi-dimensional, narratives. Furthermore, the narratives themselves, coupled with the metadata from the file system and applications, may form a meta-narrative. The literary theory of surrealism, the notion of disjointed elements, can be utilized to derive meaning from forensic evidence. This paper uses a technique known as surrealist games to illustrate the point.

Keywords: Digital forensics, narratology, surrealism

1.   Traditional Approaches to Digital Forensics

Most digital forensic examinations are done in the context of an investigation. The items that are examined are collectively referred to as "digital evidence." The Scientific Working Group on Digital Evidence (SWGDE) [21] defines digital evidence as:

> "Information of probative value stored or transmitted in binary form."

Numerous methodologies have been proposed for digital forensics (see, e.g., [5–8, 13, 17, 19]). While the methodologies differ from each other,

G. Peterson and S. Shenoi (Eds.): Advances in Digital Forensics V, IFIP AICT 306, pp. 3–15, 2009.

they have in common the goal of preserving the integrity of the original evidence and extracting information of value to the case at hand. Implicit in every methodology is the notion that the original evidence is potentially massive in size and that the "important" information is a subset, often a very small subset of the original evidence.

Numerous software tools have been developed to support digital forensic investigations: examples include EnCase, Forensic Toolkit, ProDiscover and Sleuth Kit. Most tools utilize file system structure, file type and string searches, and hash value comparisons. Novel approaches such as data mining and social network analysis have also been proposed [2, 12]. Most of these approaches rely on the technical characteristics imparted to the data by the operating system and file system. Other approaches rely on the identification of discrete bits of information whose character can be predefined, e.g., string searches and data mining. But the results, even using novel techniques such as fuzzy searches [10], have been modest at best. We believe that the effective yield – the portion of pertinent information selected by forensic techniques – is becoming not more selective, but less effective. After the potentially probative information is extracted, it becomes an analytical exercise to evaluate the information in a contextual manner. In other words, how does the information fit into the narrative of the investigation? In many cases, we try to fit the digital pieces into the framework of the case, which is, in turn, framed by the presumptive fact pattern and the elements of the law as they apply to the pattern of facts.

2.   Legal Issues

One of the hallmarks that distinguishes forensics from other applications of science is the requirement that the results be accepted as reliable evidence in a court of law. In the United States, the admission of scientific evidence is also tied to an expert witness who presents the evidence in court. An extensive body of law deals with the relationship between the evidence, the examiner and the testimony provided relative to the evidence. Two important elements are the empirical nature of the examination process and the ability of the examiner to explain the science as applied to the evidence. In the traditional model, it is important that the examination process be conducted in a "forensically sound" manner, i.e., all actions must be empirically demonstrable in both process and product.

Practitioners generally apply a scientific process to extract the evidence prior to its review for pertinence or investigative value. This tends to limit the use of novel digital forensic methodologies. If it is assumed that only the resultant facts must be demonstrably factual, then it is possible to greatly expand the potential approaches for identifying probative evidence. For sake of argument, there is no reason why we could not find the information first and then ensure its reliability by referring to an empirical process. Note, however, that we are not suggesting that this approach is a "best practice" or that it is necessarily superior to the traditional process.

The remainder of this paper explores novel digital forensic approaches.

3. Guessing the Future

Novel approaches may prove fruitful given the potential environment of the future. For the purpose of this analysis, we make the following assumptions:

- The size of the storage media to be examined will continue to grow.

- Computing devices will be used for an increasingly large number of applications.

- A greater and greater proportion of data related to a person's life will reside in electronic storage.

- We will be able to tell many more things about a person from examining electronic media.

The first assumption is highly probable; several studies have discussed this trend (see, e.g., [16]). The second assumption seems to be borne out of the phenomenon of network convergence as evidenced by devices such as the iPhone. The data related to these multipurpose devices often resides in multiple locations. Unlike the traditional model of the desktop being the primary repository of data augmented by network storage, Web 2.0 applications such as shared calendars, blogs, wikis and social networking sites store data in a slew of application servers independent of the access devices. An important characteristic of many Web 2.0 applications is "personalization," which implies that a dataset closely represents the owner.

Credit/ATM cards, access cards, toll transponders, cell phone records and network connections produce digital recordings of a person's activities. Desktops, laptops and physical media such as flash cards store our most private information. Clearly, future generations will record much more of their lives than any previous generation and the recordings will be captured and stored electronically in the web of the future. All of this should not surprise us. But how will it impact digital forensics? Perhaps a different perspective will help illuminate the issue.

4.   Hard Drive as Text

Recording the breadth of our lives in digital form can be viewed as
producing "text." Not merely numbers, letters, words, sentences, para-
graphs and pages, or even ASCII code, but text in the broadest sense of
the recording of human thought, communications and activities. One of
the many definitions for text is [22]:

> Something, such as a literary work or other cultural product, regarded
> as an object of critical analysis.

For the purposes of digital forensics, we can regard media as cultural
products or artifacts and the sum of all media associated with an individual as a collective literary work. There are many texts within this
work: letters, essays, emails, graphs, charts, diagrams, photographs, audio and video recordings. Program files are also texts, as are log files of
computer and network activity. Computer media and, by extension, all
the locations where an individual stores information constitute a digital
anthology of that person's texts.

Many kinds of texts are contained in the digital anthology. Some of
the texts are distinct, some overlapping and some redundant. The texts
are of many forms. This paper focuses on one form of text, that of the
"narrative."

5.   Narrative Theory

Part of the definition of texts that we have considered above includes
the notion of critical analysis. Students of the humanities have been
conducting critical analyses of texts for hundreds of years. This activity
is generally referred to as literary criticism. However, since the 1930s
and, especially, due to the work of Claude Levi-Strauss [14], its use
has expanded and it is now an important part of cultural anthropology.
Modern cultural anthropologists and ethnographers such as Wesch [23]
view electronic media as a cultural artifact and as a research instrument.

Mieke Bal, in her landmark text, *Narratology: Introduction to the
Theory of Narrative* [1], defines a narrative simply as "a text in which
a narrative agent tells a story." The notion of a story seems to be a
topic of literary discourse than of forensic science. However, on reflection, it should be clear that forensic scientists and investigators are well
acquainted with stories. In a very real sense, forensic scientists attempt,
through their examinations, to determine the facts and circumstances
that form the "story" of the crime. It involves the time, place, characters and action – the very things that make up a novel. Later, in
the presentation phase, the examiner writes a report that communicates
the "story" to the reader. When the examiner testifies at a trial, the

testimony conveys both the examiner's story and the story of the crime. In a real sense, forensic science is not about isolated, discrete facts, but a storytelling that communicates meaning. In our experience, forensic clients mostly ask questions about meaning, not about facts. Forensic examiners are often well equipped with facts but are not equipped to handle the meaning of the evidence.

Literary criticism is about the search for meaning. Scholars have for centuries been examining the process and the practice of human communication [18]. Many scholars have examined and continue to examine how meaning is extracted from texts. For the purposes of this paper, it would not be productive to discuss the multitude of critical approaches and schools of thought concerning the narrative. Instead, we will explore one methodology that might provide insight into how theories developed for literary criticism could be of value in a digital forensic setting.

6. Surrealism

Surrealism developed out of the despair of World War I and the rise of Dadaism. This philosophy sought new sources of inspiration from the world and the artist's mind. In his 1924 book, *Les Manifestes du Surrealisme*, Andre Breton [4] defined surrealism as:

> "Psychic automatism in its pure state, by which one proposes to express – verbally, by means of the written word, or in any other manner – the actual functioning of thought. Dictated by the thought, in the absence of any control exercised by reason, exempt from any aesthetic or moral concern."

While Breton studied psychiatry, he was not a Freudian psychoanalyst. Rather, he was trained in the French system of "dynamic psychiatry," which focused more on the elimination of the conscious than on the "unconscious" in the Freudian sense. Gibson [9] points out that Breton uses the term "depths of the mind" as opposed to the unconscious in *Les Manifestes du Surrealisme*. The significance of this concept will become apparent later in this exposition.

The initial focus of surrealism was the written word, although in time practically every form of art developed a surrealist school. The early stages of surrealism were characterized by the juxtaposition of seemingly disparate items, situations and ideas. Later, others, such as Benjamin [3], began to integrate the use of montage as a surrealist technique. In montage, isolated parts of a whole, such as sections of a photograph or portions of a projected film, were viewed in isolation. Surrealists would even use techniques such as viewing films through their fingers or a piece of cloth to disrupt the flow of the narrative [11]. They quickly discovered that these isolated pieces when placed together created a form of narrative. The technique was seen as a method for creating art and for encouraging creativity. Eventually, it became both work and pleasure – when games were developed based on the notion of montage.

7.   Surrealist Games

Surrealist games appealed to the adherents of surrealism for a number of reasons. Jean-Louis Bedouin described them as "a way of being serious without the worry of seeming so" [20]. It was the "automatic" aspect of the games that acted to suppress the conscious and free the unconscious, which it was hoped, would fuel creativity. Meanwhile, the "play" aspect served to minimize the motivation for conscious intervention in the resulting product [20].

Two popular surrealist games are Exquisite Corpse and Irrational Extension. Exquisite Corpse is a parlor game in which each player writes down a word from an assigned part of speech. These words are then combined into a phrase or metaphor. According to Ray [20]:

> "Since the philosophy of science has shown that all knowledge systems rest on a few basic metaphors, and that a new paradigm always proposes a metaphoric shift, this game might have more profound consequences than at first appears."

Irrational Extension games take a different tack. A movie is chosen as the "source" of the answers to questions posed by the person running the game. The questions are designed to be outside the scope of the source material, that is, they are designed so that there is no correct or even logical answer. The players are free to provide answers without the requirement of factual support. In addition to being humorous, these games point out holes or gaps in a narrative. By finding the questions not answered by a film, we "see" the things that our minds were likely thinking, but that were suppressed by our consciousness. Ray [20] likens this process to Freud's discovery that "resistance and repression were essential to the diagnostic process: they, in fact, pointed directly to the determining areas of a patient's experience, those leading to his symptoms."

Dove, a student of Ray, created a variant of the Irrational Extension game. In his version, the experimenter selects three "narratively important shots" (frames) and three randomly selected insignificant shots from a movie unfamiliar to the subject. All six are then presented to the subject in random order. The subject is asked a series of questions after each frame. Some of the questions are factual while others are speculative. According to Ray [20], this experiment "encourage[d] a sensitivity both to meanings communicated stereotypically and to those unconsciously." The experiment often produced unexpected results – the subject was

Figure 1. Photographs of women used in the experiment.

able to correctly identify important elements of the narrative as well as recognize important symbolism in the images.

Dove's game brings us full circle in that the power of the subconscious is used for cognition. On reflection, this should not be surprising. We are exposed to literally thousands of discrete pieces of information in our daily lives. If you were to look up from this page, you would be faced with myriad objects, sensations, responses, relationships and emotions. If you look back down, you will consciously only remember a small number of "facts," but your unconscious mind will use some of the information to contextualize the "facts;" the rest will evaporate. In a sense, we proceed through life as a series of "shots," each of which is interpreted and stored as a subset of the total image. According to Hammond [11]:

> "Stripped of their causal relations in the film, a rapid-fire of reported images emphasizes their latent content, their capacity to signify."

To test this notion, we performed an experiment in a graduate class. The class was divided into two groups, each of which was presented with one of two photographs (Figure 1). The photograph on the left, taken in Greek Cyprus, features an elderly widow dressed in the traditional black garb. The one on the right, showing a woman wearing a full-length abaya and a burqa, was taken in Alexandria, Egypt.

The students were asked five questions in the following order:

(a) What is this woman thinking?

(b) What is the next thing she will do?

(c) Why will she do it?

(d) Where is she?

(e) Why is she here?

The responses to the first three questions were predictably irrelevant as there is no information to support any objective conclusion. The answers to Questions (b) and (c) were determinable by the photographer, but not the students as they were not present at the scene. But the answers to Question (d) were objectively supportable. However, the students' responses were invariably incorrect because they did not have sufficiently detailed knowledge of the cultural contexts to answer Question (d) correctly. Since the students' responses to Question (e) were generally predicated upon their responses to (d), their responses to (e) were also incorrect.

This experiment is clearly not empirical. Rather, we wished to explore with the students how much we "know," how much we construct and when we project into the scene. This experiment seems to suggest that cognition, even subconscious cognition, requires some contextual information. The relative success of Dove's experiment may lie in his familiarity with plots, cinematic styles and history. The failure of our experiment may be due to a lack of geographic, cultural and religious background information. An additional issue may be that films are inherently narrative (a fact known to both Dove and his subjects) while photographs are not necessarily narrative. Likewise, the use of multiple "shots" in Dove's experiment provides more raw material for the subconscious.

8.   Applying Surrealist Games

So how can such a seemingly unscientific methodology be useful in forensic science? Surrealists seek to disrupt the narrative to understand the meaning of the constituent elements of a work. In order to do so, they use techniques that eliminate the narrative and suppress the context. In forensics, we do the opposite – we attempt to use the data points to discover a context and document a narrative. However, two elements of surrealist games can be exploited for forensic use: narrative context and montage.

One of the things that makes the Exquisite Corpse and Irrational Extension games and Dove's extension somewhat effective is that there is a logical progression from structure to context. In an Exquisite Corpse game, the grammatical parts of speech that are to be contributed are specified. The context and any diegesis are prevented from being provided by the secrecy used to submit the words. In our view, individuals often have an internal consistency in their submissions, which may indicate an attempt at providing a personal context. Moreover, their responses tend to have a theme.

In an Irrational Extension game, the plot and the characters present a structure and a basic context. According to Ray [20]:

> "We think that after nearly two hours with the Smith family, or fifteen movies with the Hardys, we must know everything about them and their house and their neighborhood."

In a film or other narrative, the characters are introduced, the scenes unfold with all of their subtext and relationships develop. The reader or viewer tries to fill in the blanks to create a "complete" understanding. In cinema, the story is never complete, as historian Natalie Davis was told concerning the "Camel Principle." It is not necessary to include a multitude of details to convince the audience that the scene is in a desert; the mere presence of a camel will suffice [20].

Life is fuller and much more complex. There are many more "data points" with which to complete the narrative of even one person's life. In the context of a real person with a real life, that person has a complex web of complete narratives with subplots. In fact, the problem is not a lack of data and context, it is limiting it to a manageable amount that provides an accurate picture of the individual.

In a criminal investigation, we seek to identify all the data points that are pertinent to the "subplot" of criminal activity. The person has a "complete life," only some of which is criminal in nature. If we were to surveil a criminal suspect around-the-clock, we would see that most of his activities would not be relevant to the criminal case.

A criminal who uses a computer for a length of time records a great deal of his complete narrative (life) and likely a good part of the criminal subplot on his computer hard drive. Examination of the hard drive will allow for a complete – in the sense that it exists on the hard drive – narrative of the person and his activities. It is the task of the examiner to identify and extract the portion that documents the criminal behavior. For the purposes of this discussion, the narrative and context of the criminal subplot can be developed externally through a normal investigation and/or internally through an examination of the material contained within the computer's storage (hard drive).

The second aspect of surrealist games that can be applied to digital forensics is the notion of montage. As discussed above, our own recollection of our personal narrative is a series of memories (data points) that collectively represent our subjective history. Our lives can be compared to a series of snapshots and video clips. Cinema intentionally creates montages of sequenced scenes, which are designed to convey a narrative. Photographs are frozen moments in time, in effect, a time capsule.

A digital forensic examination report is very similar because it contains selected data points that are arranged into a narrative. This narrative can be organized chronologically, topically or based on the structure of the examined media.

Since all the material on a hard drive is not included in the report, the "completeness" of the narrative is subjective and its effectiveness is measured against its "centrality" to the narrative or its pertinence to the case.

We can make use of these two aspects in the following way. We can construct a narrative externally via a traditional investigation. In this instance, we will be developing the context for the examination of the hard drive. Alternatively, we can attempt to construct the narrative and context solely by examining the content of the hard drive and any other stored electronic evidence. This, in effect, determines what the user was doing with the computer.

Next, we can select a set of data points from the evidence. Our selection of data points (files, emails, etc.) can be used to populate a narrative and/or to develop the context. The selection of these data points can be random or constrained. If it is random, it is reasonable to conclude that a larger number of data points will be necessary. If it is constrained, the effectiveness of the constraints will be a function of the centrality to the narrative. In other words, how likely is it that the constrained material contains pertinent information?

We use correspondence from the author's university email account as an example. On March 20, 2008 there were 238 emails in the account that were received during the period, March 1, 2008 through March 20, 2008. If you were to review the correspondence, you would conclude that the emails pertained to communications between the author and his superiors, peers and students. You would infer that the author was teaching two courses, enrolled as a student in two others, and active in several program committees and editorial boards. You would know about the author's research and outreach projects. You would even find a few emails from friends. The email correspondence provides a fairly complete picture of the author's professional life, since the author essentially limits the use of his university email account to professional purposes.

What if you only read eight out of every ten emails? There is a good chance that, while you might miss a few details, the narrative would still be substantially complete. Others might wish to challenge that assertion, but emails are often a series of communications and often quote previous text. As a result, the loss of 20% of the corpus is not significant; much of the information is replicated in other emails. Further, many of the "missing" details can be surmised from a careful reading of the remaining material, much like the subconscious information presented in film. We have, in effect, a "montage of data" with sufficient context to correctly answer at least some of the pertinent questions.

But would you have any useful information if you were to read only 1% of the email corpus? Unless the context is thoroughly known and the place of the data in the context is defined, it is unlikely you would have useful information. Somewhere between the 80% montage and the 1% montage, there is a sliding scale or, perhaps, a polynomial expression, that determines when the information gathered is useful. Dove's experiment demonstrated that accuracy can be attained if contextually rich sets of information are used to form even a small montage. As Benjamin [3] states:

> "In the fields with which we are concerned, knowledge comes only in flashes. The text is the thunder rolling long afterward."

9.   Conclusions

At one level, we may ask how this is any different from the way investigators already conduct the analytical aspect of forensic examinations. The answer may well be that we are consciously or subconsciously operating in this fashion. If so, surrealism, as explored in this paper, confirms this methodology as cognitively legitimate. It would benefit the digital forensic community to utilize these insights in order to perform examinations and analyses more effectively and efficiently.

The core concepts of narrative and montage are powerful tools of cognition, but they require context. Designing a system that exploits these concepts would require at least three things. First, the electronic data should be parsed into narratives. Second, these narratives and the investigative context should be coded to enable computational solutions. Third, it is necessary to understand how differing levels of narrative completeness impact the ability to make objective, sound conclusions. These requirements are not trivial and require very different approaches from those that have traditionally been used in computer science research.

The notion of utilizing surrealist gaming techniques as a digital forensic research method is not as farfetched as it might appear. One of the key issues is how to incorporate the concept of cognition into the mechanistic process of extracting information from a hard drive. Our understanding of that cognitive process currently lacks rigor. Surrealist techniques do not explain cognition, rather, they provide us with another way to "read" the hard drive. The combination of narrative and montage can inspire a new genre of powerful digital forensic tools. We must be mindful, however, of Adorno's criticism of Benjamin [15]:

> "Your study is located at the crossroad of magic and positivism."

References

[1] M. Bal, *Narratology: Introduction to the Theory of Narrative*, University of Toronto Press, Toronto, Canada, 1997.

[2] N. Beebe and J. Clark, Dealing with terabyte data sets in digital investigations, in *Advances in Digital Forensics*, M. Pollitt and S. Shenoi (Eds.), Springer, Boston, Massachusetts, pp. 3–16, 2005.

[3] W. Benjamin, Theoretics of knowledge; theory of progress, *Philosophical Forum*, vol. 15(1-2), pp. 1–40, 1984.

[4] A. Breton, *Les Manifestes du Surrealisme*, Jean-Jacques Pauvert, Paris, France, 1972.

[5] S. Bunting and S. Anson, *Mastering Windows Network Forensics and Investigation*, Sybex, Alameda, California, 2007.

[6] B. Carrier, *File System Forensic Analysis*, Addison-Wesley, Boston, Massachusetts, 2005.

[7] H. Carvey, *Windows Forensic Analysis*, Syngress, Rockland, Massachusetts, 2007.

[8] E. Casey, *Digital Evidence and Computer Crime*, Academic Press, Boston, Massachusetts, 2004.

[9] J. Gibson, Surrealism before Freud: Dynamic psychiatry's "simple recording instrument," *Art Journal*, vol. 46(1), pp. 56–60, 1987.

[10] J. Guan, D. Liu and T. Wang, Applications of fuzzy data mining methods for intrusion detection systems, *Proceedings of the International Conference on Computational Science and its Applications*, pp. 706–714, 2004.

[11] P. Hammond, *The Shadow and its Shadow: Surrealist Writings on the Cinema*, City Lights Books, San Francisco, California, 2000.

[12] M. Hoeschele and M. Rogers, Detecting social engineering, in *Advances in Digital Forensics*, M. Pollitt and S. Shenoi (Eds.), Springer, Boston, Massachusetts, pp. 67–77, 2005.

[13] W. Kruse and J. Heiser, *Computer Forensics: Incident Response Essentials*, Addison-Wesley, Boston, Massachusetts, 2001.

[14] E. Leach, The social theory of Claude Levi-Strauss, *British Journal of Sociology*, vol. 33(1), pp. 148–149, 1982.

[15] H. Lonitz (Ed.), *Theodor W. Adorno and Walter Benjamin: The Complete Correspondence, 1928-1940*, Harvard University Press, Cambridge, Massachusetts, 1999.

[16] P. Lyman and H. Varian, How much information? University of California, Berkeley, California (www2.sims.berkeley.edu/research/projects/how-much-info-2003), 2003.

[17] A. Marcella and D. Menendez, *Cyber Forensics: A Field Manual for Collecting, Examining and Preserving Evidence of Computer Crimes*, Auerbach Publications, Boca Raton, Florida, 2007.

[18] W. Ong, *Orality and Literacy*, Routledge, New York, 2002.

[19] M. Pollitt and S. Shenoi (Eds.), *Advances in Digital Forensics*, Springer, Boston, Massachusetts, 2005.

[20] R. Ray, *The Avant-Garde Finds Andy Hardy*, Harvard University Press, Cambridge, Massachusetts, 1995.

[21] Scientific Working Groups on Digital Evidence and Imaging Technology, SWGDE and SWGIT Digital and Multimedia Evidence Glossary (www.swgde.org/documents/swgde2008/SWGDE_SWGITGlossaryV2.2.pdf), 2007.

[22] TheFreeDictionary.com, text (www.thefreedictionary.com/text).

[23] M. Wesch, Digital ethnography (mediatedcultures.net/about.htm).

Chapter 2

DIGITAL FORENSIC RESEARCH: THE GOOD, THE BAD AND THE UNADDRESSED

Nicole Beebe

Abstract   Digital forensics is a relatively new scientific discipline, but one that has matured greatly over the past decade. In any field of human endeavor, it is important to periodically pause and review the state of the discipline. This paper examines where the discipline of digital forensics is at this point in time and what has been accomplished in order to critically analyze what has been done well and what ought to be done better. The paper also takes stock of what is known, what is not known and what needs to be known. It is a compilation of the author's opinion and the viewpoints of twenty-one other practitioners and researchers, many of whom are leaders in the field. In synthesizing these professional opinions, several consensus views emerge that provide valuable insights into the "state of the discipline."

Keywords: Digital forensic research, evaluation, future research areas

1.   Introduction

Digital forensics is defined as: "[t]he use of scientifically derived and proven methods toward the preservation, collection, validation, identification, analysis, interpretation, documentation and presentation of digital evidence derived from digital sources for the purpose of facilitation or furthering the reconstruction of events found to be criminal, or helping to anticipate unauthorized actions shown to be disruptive to planned operations" [46]. This paper presents the results of a "state of the discipline" examination of digital forensics. The focus is on digital forensic research – systematic, scientific inquiries of facts, theories and problems related to digital forensics. In particular, the paper character-

G. Peterson and S. Shenoi (Eds.): Advances in Digital Forensics V, IFIP AICT 306, pp. 17–36, 2009.

izes the current body of knowledge in digital forensics and evaluates its rigor with the goal of setting a course for future digital forensic research.

This evaluation of the current state of digital forensics attempts to convey the perspectives of researchers and practitioners. In addition to the author, the views expressed in this paper are drawn from twenty-one researchers and practitioners, many of whom are leaders in the field. Researchers were asked to critically examine the collective contribution of the research community to the current body of knowledge and to identify the most pressing research questions in digital forensics. Practitioners were queried about current and previous research activities, the contribution of these activities to their real-world experiences, and pressing research needs.

The remainder of the paper is organized as follows. First, we discuss what is collectively seen as "good" in the field, outlining the notable research contributions over the past decade. Next, we discuss the "bad" – what needs to be improved as far as digital forensic research is concerned. Finally, we attempt to set a course for future digital forensic research by discussing four major themes and several individual research topics that demand investigation. A brief literature review of the research themes and topics is presented to assist individuals who are interested in embarking on research in these areas.

2. The Good

All the respondents felt that there was unequivocal improvement in the prominence and value of digital evidence in investigations. It is now mainstream knowledge that the digital footprints that remain after interactions with computers and networks are significant and probative. Digital forensics was once a niche science that was leveraged primarily in support of criminal investigations, and digital forensic services were utilized only during the late stages of investigations after much of the digital evidence was already spoiled. Now, digital forensic services are sought right at the beginning of all types of investigations – criminal, civil, military and corporate. Even popular crime shows and novels regularly incorporate digital evidence in their story lines.

The increased public awareness of digital evidence says nothing about the state of digital forensics as a science. Indeed, the awareness of the need to collect and analyze digital evidence does not necessarily translate to scientific theory, scientific processes and scientifically derived knowledge. The traditional forensic sciences (e.g., serology, toxicology and ballistics) emerged out of academic research, enabling science to precede forensic science applications, as it should. Digital forensics, however, emerged out of the practitioner community – computer crime investigators and digital forensic tool developers seeking solutions to real-world problems [47]. While these efforts have produced a great amount of factual knowledge and several commonly accepted processes and hardware and software tools, many experts concede that the scientific method did not underlie much of early digital forensic research.

The call for a more scientific approach to digital forensic research is not new. The first Digital Forensic Research Workshop (DFRWS 2001) convened more than 50 researchers, investigators and analysts "...to establish a research community that would apply the scientific method in finding focused near-term solutions driven by practitioner requirements and addressing longer term needs, considering, but not constrained by current paradigms" [46].

The prevailing sentiment is that the scientific foundation of digital forensics has been strengthened. A few examples highlight this point. The Scientific Working Group on Digital Evidence (SWGDE) has released several documents since 1999 concerning digital forensic standards, best practices and testing and validation processes. In 2001, the U.S. National Institute of Standards and Technology (NIST) started the Computer Forensic Tool Testing (CFTT) Project and has since established and executed validation test protocols for several digital forensic tools. DFRWS 2002 focused on scientific standards and methods. DFRWS 2003 hosted a plenary session emphasizing the need for a scientific foundation for digital forensic research. DFRWS 2004 featured several presentations defining the digital forensic process. In short, the field has experienced considerable progress in formalizing, standardizing and formulating digital forensic processes and approaches. The respondents also felt that the research community was showing signs of a stronger scientific underpinning as evidenced by the publication of research in mainstream computer science, information systems and engineering journals.

To date, research questions have largely centered on the "archaeology" of digital artifacts. Carrier [12] observes that digital forensic artifacts are a function of the physical media, operating system, file system and user-level applications – that each impacts what digital evidence is created and left behind. Like archaeologists who seek to understand past human behavior by studying artifacts, digital forensic investigators seek to understand past behavior in the digital realm by studying digital artifacts. Because digital forensic research during the past decade has focused on the identification, excavation and examination of digital artifacts, there is now a relatively solid understanding of what digital artifacts exist, where they exist, why they exist and how to recover them (relative to commonly-used operating/file systems and software applications). To its credit, the digital forensic research community has done a good job sharing this knowledge with other academic disciplines (e.g., computer science, information systems, engineering and criminal justice) as well as with the practitioner community (law enforcement, private-sector practitioners and e-discovery specialists).

Digital forensic research has profited from "digital forensic challenges" designed to stimulate scientific inquiry and the development of innovative tools and analytical methodologies. Examples are the annual digital forensic challenges sponsored by DFRWS (since 2005) and the U.S. DoD Cyber Crime Center (since 2006). Two research areas that have experienced significant growth as a result of the challenges are data carving and memory analysis. Other areas that have benefited include steganography, encryption and image identification (especially distinguishing between real and computer-generated or computer-altered images).

However, the digital forensic challenges may have shifted research attention away from the response and data collection phases to the analysis phase. Many of the respondents opined that digital forensic research initially focused its attention on response and data collection. As a result, robust hardware write blockers became widely available; live response processes, tools and methodologies that minimized the digital evidence footprint were developed; and commonly accepted acquisition policies and procedures emerged. One might argue that the research community has marched along the digital forensic process: Preparation → Response → Collection → Analysis → Presentation → Incident Closure [4]. Many research questions pertaining to the response and collection phases are still unanswered, it is just that the research community now sees the most pressing questions as residing in the analysis phase.

3.   The Bad

Interestingly, several of the key successes of digital forensic research follow directly into the discussion of "The Bad." Take, for example, acquisition process standardization and formalization. Several of the respondents suggested that the digital forensic community has almost hyper-formalized processes and approaches, especially with respect to the response and data collection phases. Some would argue this point, citing the fact that there is no single, universal standard for digital evidence collection. Many organizations have their own standards and guidelines for data collection. Further, it can be argued that these documents constitute high-level, work-flow guidance rather than proscriptive checklists. Thus, at first glance, one can easily make an argument against the allegation of hyper-formalization.

The argument that digital forensic processes are hyper-formalized centers on the fact that the evidentiary principles established over the years cannot be attained under certain circumstances. Consider the evidentiary principles of integrity and completeness. The digital forensic community has worked hard to get the judiciary to understand that the right way to respond and collect digital evidence does not alter the evidence in any way and obtains all the evidence. The problem is that the changing technological landscape often necessitates a different approach – one where evidence will be altered (albeit minimally and in a deterministic manner) and where not all the evidence can be seized. Modern digital crime scenes frequently involve multi-terabyte data stores, mission-critical systems that cannot be taken offline for imaging, ubiquitous sources of volatile data, and enterprise-level and/or complex incidents in which the scope and location of digital evidence are difficult to ascertain. Many organizational standards and guidelines fail to address response and data acquisition in such circumstances; they often fail to facilitate proper decision-making in the light of unexpected digital circumstances; and they often present evidentiary principles as "rules," leaving little room for improvisation.

One of the successes identified in the previous section was the collective ability to archaeologically identify, excavate and examine digital artifacts. The problem, however, is that knowledge and expertise are heavily biased toward Windows, and to a lesser extent, standard Linux distributions. The FAT12/16/32, NTFS and EXT2/3 file systems, the operating systems that implement them (Windows9X/ME/NT/XP/Vista and various Linux distributions), and common user applications installed on them (Microsoft Internet Explorer and Outlook, Mozilla Firefox and Thunderbird, etc.) have been well studied. Researchers have paid insufficient attention to other operating systems, file systems and user applications, especially in the light of current market trends.

The market share enjoyed by Microsoft operating systems has decreased from 91.8% in May 2008 to 88.1% in March 2009 [43], due in large part to Apple's Mac OS X and its portable device operating systems. "Mac forensics" (e.g., HFS+ file system forensics) has received increased research attention, but more efforts are needed. New file systems, such as ZFS by Sun Microsystems, have received little attention. UFS and ReiserFS are also examples of file systems that deserve to be the focus of more research.

The digital forensic research community must also challenge itself by raising the standards for rigor and relevance of research in digital forensics. In years past, the challenge for the community was limited knowledge, skills and research experience. Despite backgrounds in computer science and information systems, and in some cases, real-world digital forensic training and case experience, members of the research community still had to overcome a steep learning curve with respect to the core body of knowledge in digital forensics (if, in fact, such a core ever existed). This unfortunately led to lower standards for scientific rigor and relevance in digital forensic research compared with other traditional fields of research.

The problem of rigor and relevance in digital forensic research is exacerbated by two publication dilemmas. First, the receptivity of mainstream scientific journals to digital forensic research is nascent. Second, peer reviewed, scientific journals dedicated entirely to digital forensics are relatively new. Journals typically take time to achieve high quality standards (a function of increasing readership and decreasing acceptance rates over time), and the discipline of digital forensics is simply not there yet. A survey of journals dedicated to digital forensics reveals that the ISI impact factors, circulation rates and acceptance rates are below par (but they are improving). This problem will subside as time passes and the discipline grows and matures.

Regardless of the mitigating circumstances noted above, the research community must regulate itself and "raise the bar." It should ensure that every research publication makes a clear (even if only incremental) contribution to the body of knowledge. This necessitates exhaustive literature reviews. More importantly, the work should be scientifically sound and comparable in rigor to research efforts in more established scientific disciplines.

It is also important to ensure that digital forensic research is relevant to the practitioner community. One strategy is to better address the problem phrased by Pollitt as "data glut, knowledge famine." Investigators usually do not lack data, but they often struggle with transforming the data into investigative knowledge. Research should strive to minimize noise and maximize contextual information, thereby converting data to investigative knowledge (Figure 1). The second strategy is for digital forensic research to facilitate tool development (at least to some degree). While some believe it is the responsibility of the digital forensic research community to develop usable tools, many of the respondents disagreed with this assertion. Nonetheless, it is important to ensure that the research is accessible and communicated to digital forensic tool developers, so that key contributions to the body of knowledge are placed in the hands of practitioners.

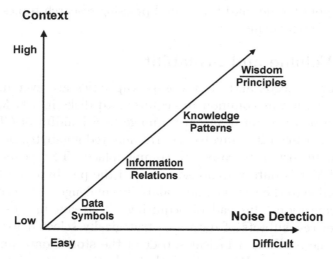

Figure 1. Knowledge management understanding hierarchy [44].

All things considered, "The Bad" is not that horrible. Rather, it calls for a reorientation and introspection by the digital forensic research community. As the research community approaches the apex of the learning curve, it should return to its scientific roots. It should reach deep into the relevant reference disciplines, build theory via strong analytics and methodological rigor, conduct scientifically sound experiments, and demand that all publications make clear contributions to the body of knowledge in digital forensics. The digital forensic community as a whole must achieve an awareness of the changing technological landscape that goes beyond practice. Indeed, researchers should be the ones advising practitioners about technologies on the horizon, not vice versa. Researchers should be the ones driving tool development based on their studies of archaeological artifacts, human analytical approaches and computational search and extraction algorithms, rather than having the tools drive digital forensic analytical approaches, processes and outcomes. Overall, digital forensic research must become more rigorous, relevant and forward thinking.

4. The Unaddressed

Having reviewed the current state of digital forensics, it is important to identify strategic directions for research in the field. Our study reveals four major themes: (i) volume and scalability, (ii) intelligent analytical approaches, (iii) digital forensics in and of non-standard computing environments, and (iv) forensic tool development. The following sections discuss the four themes and list several pressing research topics after the discussion of each theme.

4.1   Volume and Scalability

Data storage needs and data storage capacities are ever increasing. Ten years ago, it was common to acquire hard disks in 700 MB image segments in order to burn an entire image to a handful of CD-ROMs. Now, "small" cases often involve several hundred gigabytes of data and multi-terabyte corporate cases are commonplace. Two years ago, the size of Wal-Mart's data warehouse exceeded the petabyte mark [75].

One solution to the volume and scalability challenge is selective digital forensic acquisition. Instead of acquiring bit-stream images of entire physical devices, subsets of data are strategically selected for imaging. Typically, the result is a logical subset of the stored data and not all logical data at that. We contend that selective acquisition can and should include certain portions of allocated and unallocated space, but admittedly, research is needed to facilitate such acquisitions (especially related to the decision making process that would identify the data to be selectively acquired).

Research on selective, intelligent acquisition includes using digital evidence bags [71, 72] and risk sensitive digital evidence collection [32]. Digital evidence bags are designed to store provenance information related to the data collected via selective acquisition. This type of approach is important because, when acquiring subsets of data from disparate sources, the source and contextual data (i.e., the physical device and the subset of data that is not acquired) are no longer implicitly available and must be explicitly retained. Furthermore, any explicitly retained information can and should be managed in order to contribute knowledge to the analytical process. Risk sensitive collection provides a framework for allowing cost-benefit considerations to drive the selection process, considering costs and benefits to the investigating and data-owning entities.

Another solution to the volume and scalability challenge is to utilize more effective and efficient computational and analytical approaches. Research focused on improving the efficiency and effectiveness of digital forensic analyses include distributed analytical processing [56], data-mining-based search processes [5], file classification to aid analysis [60], self-organizing neural networks for thematically clustering string search results [6], and massive threading via graphical processing units (GPUs) [38]. Researchers have also investigated network-based architectures and virtualization infrastructures to facilitate large evidence stores, case and digital asset management systems, and collaborative, geographically distributed analysis [19, 20].

Clearly, more research is needed to address volume and scalability issues. The following research questions are proffered for further inquiry by the research community.

- How can decision support systems be extended to the digital forensics realm to aid selective and intelligent acquisition?

- What are the dimensions of the selective acquisition decision making process and how do they differ from other processes where decision support systems are applied?

- How can data warehousing and associated information retrieval and data mining research be extended to digital forensics? Beebe and Clark [5] discuss how data mining can be extended to digital forensics, but more research is needed. Which information retrieval and data mining approaches and algorithms can be extended to address specific analytical problems?

- Since information retrieval and data mining approaches are typically designed to handle logical and relatively homogeneous data sets, what adaptations are necessary to deal with physical, highly heterogeneous data sets?

- Link analysis research has been extended to non-digital investigations in the form of crime network identification and analysis, but how can similar research be extended to digital investigations and data sets? Can this research help characterize the relationships between digital events (chronological or otherwise) and digital data?

- How do investigators search and analyze data? What cognitive processing models are involved? How do these compare with other human information processing, knowledge generation and decision making processes?

- How does and/or should Simon's concept of "satisficing" [66] extend to digital forensic investigations?

- Why is digital forensic software development lagging hardware advances in the areas of large-scale multi-threading and massive parallelism? Are there unique characteristics about data and information processing tasks in a digital forensic environment that make it more difficult or necessitate different development or engineering approaches?

4.2   Intelligent Analytical Approaches

The second major research theme is intelligent analytical approaches. Several respondents felt that computational approaches for searching, retrieving and analyzing digital evidence are unnecessarily simplistic. Current approaches largely rely on: (i) literal string searching (i.e., non-**grep** string searches for text and file signatures), (ii) simple pattern matching (i.e., **grep** searches), (iii) indexing data to speed up searching and matching, (iv) hash analyses, and (v) logical level file reviews (i.e., log analysis, registry analysis, Internet browser file parsing, viewing allocated files, etc.). There are two problems associated with these approaches: underutilization of available computational power and high information retrieval overhead.

Current information retrieval and analytical approaches underutilize available computational power. Many forensic search processes require large amounts of processing time and researchers continue to seek ways to conduct searches and analyze data more quickly [38, 56]. However, the amount of time required to conduct byte-by-byte matching or full-text indexing is not the issue. The point is that high-end, user-class computing platforms (akin to typical digital forensic workstations) can handle intelligent search, retrieval and analytical algorithms that are much more advanced than literal string searches and simple pattern matching. Advanced algorithms already exist and are the result of longstanding research efforts in artificial intelligence, information science, data mining and information retrieval.

Current search and analysis approaches also have significant information retrieval overhead. In addition to the computational time required to execute a search, the overhead includes the human information processing time spent to review hits that are not relevant to the investigative objectives (i.e., false positives in the investigative sense).

The classic precision-recall trade-off dilemma is that as recall increases, the query precision decreases. Since digital forensics seeks recall rates at or near 100%, query precision is usually low. The heterogeneity of data sources during physical-level forensic analysis intensifies the problem. Note that traditional query recall targets could be reconsidered in the light of legal sufficiency and Simon's notion of satisficing [66].

The cost of human analytical time spent sifting through non-relevant search hits is a significant issue. Skilled investigators are in limited quantity, highly paid and often face large case backlogs. Anything that can be done to reduce the human burden should be seriously considered, even if it means increasing computational time. Trading equal human analytical time for computer processing time is a worthwhile proposition in and of itself, but we believe that the trade will seldom be equal. Extending intelligent search, retrieval and analytical algorithms will not require a one-for-one trade between human and computer time. Computational processing time will indeed increase, but it will pale in comparison with the amount and cost of human analytical time savings.

Research in intelligent analytical approaches is relatively scant. Much of the work was discussed in the context of volume and scalability challenges. "Smarter" analytical algorithms would clearly reduce information retrieval overhead. They should help investigators get to relevant data more quickly, reduce the noise investigators must wade through, and help transform data into information and investigative knowledge.

In addition to improving analytical efficiency, intelligent analytical approaches would enhance analytical effectiveness. Research has shown that data mining algorithms can reveal data trends and information otherwise undetectable by human observation and analysis. Indeed, the increased application of artificial intelligence, information science, data mining and information retrieval algorithms to digital forensics will enable investigators to obtain unprecedented investigative knowledge.

The following topics in the area of intelligent analytical approaches should be of interest to the research community:

- Advances in the use and implementation of hashing, including the use of bloom filters to improve the efficiency of hash analyses [55], and the use of hashes as probabilistic, similarity measures instead of binary measures of identicalness [34, 57, 58].

- Using self-organizing neural networks to thematically cluster string search results and provide relevant search hits significantly faster than otherwise possible [6].

- Automated, large-scale file categorization within homogeneous file classes or file types [60].

- Statistically assessing "Trojan defense" claims [11].

- Feature-based data classification without the aid of file signatures or file metadata [64].

- Applying artificial intelligence techniques (e.g., support vector machines and neural networks) to analyze offline intrusion detection data and detect malicious network events [42].

- Using association rule mining for log data analysis and anomaly detection [1, 2].

- Using support vector machines for email attribution [21].

4.3   Non-Standard Computing Environments

The standard computing environment has long been the personal computer (desktops and laptops). Accordingly, digital forensic research has focused on acquiring and analyzing evidence from hard disk drives and memory. However, the technological landscape is changing rapidly – there is no longer a single "standard" computing environment. Small, mobile devices are ubiquitous and vary greatly (e.g., mobile phones, PDAs, multimedia players, GPS devices, gaming systems, USB thumb drives, etc.). Virtualization is widespread in personal and organizational computing infrastructures. Cloud computing is rapidly relocating digital evidence and commingling it with data from other organizations, which introduces new legal challenges. Operating systems and file systems are no longer 90% Microsoft Windows based. Investigators are encountering large numbers of custom-built digital devices. How will the digital forensic community deal with these non-standard computing environments and devices?

Researchers have made great strides in the area of small device forensics (see, e.g., [10, 13, 14, 22, 30, 33, 37, 41, 61, 67, 69, 70, 74, 76]). But much more work remains to be done given the rapid pace with which new models and devices enter the market.

Virtualization is also an area that deserves attention. Most research efforts have focused on leveraging virtualization in digital forensic analytic environments [8, 19, 49] or in educational environments [51] rather than conducting digital forensics of virtual environments. Dorn and co-workers [25] recently examined the digital forensic impact of virtual machines on their hosts. This work falls in the important area of "analysis of virtual environments" [51], which includes forensic data acquisition, virtual platform forensics and virtual introspection.

Cloud computing is another area that has received little attention. Most research is geared towards leveraging cloud computing (data center CPUs) to conduct digital forensic investigations more efficiently (see, e.g., [59]). To our knowledge, no research has been published on how cloud computing environments affect digital artifacts, and on acquisition logistics and legal issues related to cloud computing environments.

Clearly, there are benefits to be gained from researching virtualization and cloud computing as resources for completing digital forensic investigations as well as for their impact on digital forensic artifacts. To date, however, there has been little research in either direction.

Digital forensic research is also playing catch-up with non-Microsoft-based operating systems and file systems. Mac forensics is a growing field, and necessarily so, but more research is required in this area. Other important areas of research are HFS+ and ZFS file system forensics [7, 9, 18, 31, 40] and Linux and Unix system forensics [17, 26, 50].

4.4   Forensic Tool Development

The fourth and final unaddressed research theme is the design and implementation of digital forensic tools. Many of the respondents believed that current tools were somewhat limited in terms of their ease of use and software engineering.

Ease of use is a major issue. Tools must not be too technical and must have intuitive interfaces, but, at the same time, they should be customizable for use by skilled practitioners. Furthermore, the goal should be to provide information and knowledge, not merely data. This might be accomplished through data visualization, automated link analysis, cross-correlation and features for "zooming in" on information to reduce information overhead. Another approach is to shift from the tradition of presenting data hierarchically based on file system relationships to presenting data temporally. The digital forensic research community should consider, extend and adapt approaches devised by graphics and visualization and human-computer interaction researchers.

The respondents also suggested several improvements with respect software engineering. Software development must take advantage of hardware advances, including massive parallelism and streaming. Increased interoperability via standardized data (i.e., tool input/output) and API formats is needed. More operating-system-independent tools (e.g., PyFlag [16]) are required. All-in-one tools with respect to data types are needed; such tools would intelligently leverage data from static media, volatile memory/devices, network dumps, etc. It is also important to increase the automation of forensic processes.

These suggestions beg the fundamental question: Why are digital forensic tools not there yet? Is it a symptom of the relative nascence of the field and, thus, the tools? Or, are digital forensic analytical tasks fundamentally different from tasks in other domains where similar technologies work? If so, are the differences regarding the human analytical sub-task, the computational sub-task, or both? Is this even a valid research stream for digital forensic researchers, or should it be left to commercial software developers? Do these questions necessitate research, or simply awareness of the problem on the part of tool developers? In any case, these questions must be considered and collectively answered by the digital forensic community as soon as possible.

4.5   Other Important Research Topics

In addition to the four major themes, several important research topics emerged during our discussions with digital forensic researchers and practitioners. These include:

- Detection, extraction and analysis of steganographically-inserted data, particularly data inserted by non-standard stego applications (see, e.g., [3, 27, 29, 39, 52–54, 65]).

- Database forensics (see, e.g., [45]).

- Forensic processes on live and volatile sources of digital evidence, evidentiary disturbance caused by memory acquisition and live forensic analysis, and evidentiary integrity processes and standards (see, e.g., [23, 24, 28, 35, 62, 63, 68, 73]).

- Emergent metadata standards and trends (e.g., new XML Office document standards) (see, e.g., [48]).

- Investigations involving multiple, distributed systems.

- Increased insight into error rates as they pertain to digital forensic tools, processes, algorithms and approaches (especially related to Daubert standards). It is not clear if the traditional concept of error rates is appropriate or if the paradigm should be shifted to the determination and quantification of the confidence of conclusions (see, e.g., [15]).

- Formalization of the hypothesis generation and testing process in examinations.

- Experimental repeatability and comparability of scientific research findings (including the creation of common test corpora).

5.   Conclusions

Digital forensic research has experienced many successes during the past decade. The importance of digital evidence is now widely recognized and the digital forensic research community has made great strides in ensuring that "science" is emphasized in "digital forensic science." Excellent work has been accomplished with respect to identifying, excavating and examining archaeological artifacts in the digital realm, especially for common computing platforms. Also, good results have been obtained in the areas of static data acquisition, live forensics, memory acquisition and analysis, and file carving.

However, strong efforts should be directed towards four key research themes and several individual research topics. The four key themes are: (i) volume and scalability challenges, (ii) intelligent analytical approaches, (iii) digital forensics in and of non-standard computing environments, and (iv) forensic tool development. In addition to these larger themes, pressing research topics include steganography detection and analysis, database forensics, live file system acquisition and analysis, memory analysis, and solid state storage acquisition and analysis.

Acknowledgements

The following individuals have contributed to this assessment of the discipline of digital forensics: Frank Adelstein, Dave Baker, Florian Bucholz, Ovie Carroll, Eoghan Casey, DeWayne Duff, Drew Fahey, Simson Garfinkel, John Garris, Rod Gregg, Gary Kessler, Gary King, Jesse Kornblum, Russell McWhorter, Mark Pollitt, Marc Rogers, Vassil Roussev, Sujeet Shenoi, Eric Thompson, Randy Stone and Wietse Venema. Their assistance is gratefully acknowledged.

References

[1] T. Abraham and O. de Vel, Investigative profiling with computer forensic log data and association rules, *Proceedings of the IEEE International Conference on Data Mining*, pp. 11–18, 2002.

[2] T. Abraham, R. Kling and O. de Vel, Investigative profile analysis with computer forensic log data using attribute generalization, *Proceedings of the Fifteenth Australian Joint Conference on Artificial Intelligence*, 2002.

[3] K. Bailey and K. Curran, An evaluation of image based steganography methods, *International Journal of Digital Evidence*, vol. 2(2), 2003.

[4] N. Beebe and J. Clark, A hierarchical, objectives-based framework for the digital investigations process, *Digital Investigation*, vol. 2(2), pp. 147–167, 2005.

[5] N. Beebe and J. Clark, Dealing with terabyte data sets in digital investigations, in *Advances in Digital Forensics*, M. Pollitt and S. Shenoi (Eds.), Springer, Boston, Massachusetts, pp. 3–16, 2005.

[6] N. Beebe and J. Clark, Digital forensic text string searching: Improving information retrieval effectiveness by thematically clustering search results, *Digital Investigation*, vol. 4(S1), pp. 49–54, 2007.

[7] N. Beebe, S. Stacy and D. Stuckey, Digital forensic implications of ZFS, to appear in *Digital Investigation*, 2009.

[8] D. Bem and E. Huebner, Computer forensic analysis in a virtual environment, *International Journal of Digital Evidence*, vol. 6(2), 2007.

[9] A. Burghardt and A. Feldman, Using the HFS+ journal for deleted file recovery, *Digital Investigation*, vol. 5(S1), pp. 76–82, 2008.

[10] P. Burke and P. Craiger, Forensic analysis of Xbox consoles, in *Advances in Digital Forensics III*, P. Craiger and S. Shenoi (Eds.), Springer, Boston, Massachusetts, pp. 269–280, 2007.

[11] M. Carney and M. Rogers, The Trojan made me do it: A first step in statistical based computer forensics event reconstruction, *International Journal of Digital Evidence*, vol. 2(4), 2004.

[12] B. Carrier, *File System Forensic Analysis*, Addison-Wesley, Boston, Massachusetts, 2005.

[13] H. Carvey, Tracking USB storage: Analysis of Windows artifacts generated by USB storage devices, *Digital Investigation*, vol. 2(2), pp. 94–100, 2005.

[14] F. Casadei, A. Savoldi and P. Gubian, Forensics and SIM cards: An overview, *International Journal of Digital Evidence*, vol. 5(1), 2006.

[15] E. Casey, Error, uncertainty and loss in digital evidence, *International Journal of Digital Evidence*, vol. 1(2), 2002.

[16] M. Cohen, PyFlag – An advanced network forensic framework, *Digital Investigation*, vol. 5(S1), pp. 112–120, 2008.

[17] P. Craiger, Recovering digital evidence from Linux systems, in *Advances in Digital Forensics*, M. Pollitt and S. Shenoi (Eds.), Springer, Boston, Massachusetts, pp. 233–244, 2005.

[18] P. Craiger and P. Burke, Mac OS X forensics, in *Advances in Digital Forensics II*, M. Olivier and S. Shenoi (Eds.), Springer, Boston, Massachusetts, pp. 159–170, 2006.

[19] P. Craiger, P. Burke, C. Marberry and M. Pollitt, A virtual digital forensics laboratory, in *Advances in Digital Forensics IV*, I. Ray and S. Shenoi (Eds.), Springer, Boston, Massachusetts, pp. 357–365, 2008.

[20] M. Davis, G. Manes and S. Shenoi, A network-based architecture for storing digital evidence, in *Advances in Digital Forensics*, M. Pollitt and S. Shenoi (Eds.), Springer, Boston, Massachusetts, pp. 33–43, 2005.

[21] O. de Vel, A. Anderson, M. Corney and G. Mohay, Mining email content for author identification forensics, *ACM SIGMOD Record*, vol. 30(4), pp. 55–64, 2001.

[22] A. Distefano and G. Me, An overall assessment of mobile internal acquisition tool, *Digital Investigation*, vol. 5(S1), pp. 121–127, 2008.

[23] B. Dolan-Gavitt, The VAD tree: A process-eye view of physical memory, *Digital Investigation*, vol. 4(S1), pp. 62–64, 2007.

[24] B. Dolan-Gavitt, Forensic analysis of the Windows registry in memory, *Digital Investigation*, vol. 5(S1), pp. 26–32, 2008.

[25] G. Dorn, C. Marberry, S. Conrad and P. Craiger, Analyzing the impact of a virtual machine on a host machine, in *Advances in Digital Forensics V*, G. Peterson and S. Shenoi (Eds.), Springer, Heidelberg, Germany, pp. 69–81, 2009.

[26] K. Eckstein and M. Jahnke, Data hiding in journaling file systems, *Proceedings of the Fifth Digital Forensic Research Workshop*, 2005.

[27] C. Hosmer and C. Hyde, Discovering covert digital evidence, *Proceedings of the Third Digital Forensic Research Workshop*, 2003.

[28] E. Huebner, D. Bem, F. Henskens and M. Wallis, Persistent systems techniques in forensic acquisition of memory, *Digital Investigation*, vol. 4(3-4), pp. 129–137, 2007.

[29] J. Jackson, G. Gunsch, R. Claypoole and G. Lamont, Blind steganography detection using a computational immune system approach: A proposal, *Proceedings of the Second Digital Forensic Research Workshop*, 2002.

[30] W. Jansen and R. Ayers, An overview and analysis of PDA forensic tools, *Digital Investigation*, vol. 2(2), pp. 120–132, 2005.

[31] R. Joyce, J. Powers and F. Adelstein, MEGA: A tool for Mac OS X operating system and application forensics, *Digital Investigation*, vol. 5(S1), pp. 83–90, 2008.

[32] E. Kenneally and C. Brown, Risk sensitive digital evidence collection, *Digital Investigation*, vol. 2(2), pp. 101–119, 2005.

[33] M. Kiley, T. Shinbara and M. Rogers, iPod forensics update, *International Journal of Digital Evidence*, vol. 6(1), 2007.

[34] J. Kornblum, Identifying almost identical files using context triggered piecewise hashing, *Digital Investigation*, vol. 3(S1), pp. 91–97, 2006.

[35] J. Kornblum, Using every part of the buffalo in Windows memory analysis, *Digital Investigation*, vol. 4(1), pp. 24–29, 2007.

[36] G. Kowalski and M. Maybury, *Information Storage and Retrieval Systems: Theory and Implementation*, Kluwer, Norwell, Massachusetts, 2000.

[37] C. Marsico and M. Rogers, iPod forensics, *International Journal of Digital Evidence*, vol. 4(2), 2005.

[38] L. Marziale, G. Richard and V. Roussev, Massive threading: Using GPUs to increase the performance of digital forensic tools, *Digital Investigation*, vol. 4(S1), pp. 73–81, 2007.

[39] B. McBride, G. Peterson and S. Gustafson, A new blind method for detecting novel steganography, *Digital Investigation*, vol. 2(1), pp. 50–70, 2005.

[40] K. McDonald, To image a Macintosh, *Digital Investigation*, vol. 2(3), pp. 175–179, 2005.

[41] B. Mellars, Forensic examination of mobile phones, *Digital Investigation*, vol. 1(4), pp. 266–272, 2004.

[42] S. Mukkamala and A. Sung, Identifying significant features for network forensic analysis using artificial intelligence techniques, *International Journal of Digital Evidence*, vol. 1(4), 2003.

[43] Net Applications, Global Market Share Statistics, Aliso Viejo, California (marketshare.hitslink.com), April 9, 2009.

[44] J. Nunamaker, N. Romano and R. Briggs, A framework for collaboration and knowledge management, *Proceedings of the Thirty-Fourth Hawaii International Conference on System Sciences*, 2001.

[45] M. Olivier, On metadata context in database forensics, *Digital Investigation*, vol. 5(3-4), pp. 115–123, 2009.

[46] G. Palmer, A Road Map for Digital Forensic Research, DFRWS Technical Report, DTR-T001-01 Final, Air Force Research Laboratory, Rome, New York, 2001.

[47] G. Palmer, Forensic analysis in the digital world, *International Journal of Digital Evidence*, vol. 1(1), 2002.

[48] B. Park, J. Park and S. Lee, Data concealment and detection in Microsoft Office 2007 files, *Digital Investigation*, vol. 5(3-4), pp. 104–114, 2009.

[49] M. Penhallurick, Methodologies for the use of VMware to boot cloned/mounted subject hard disks, *Digital Investigation*, vol. 2(3), pp. 209–222, 2005.

[50] S. Piper, M. Davis, G. Manes and S. Shenoi, Detecting hidden data in EXT2/EXT3 file systems, in *Advances in Digital Forensics*, M. Pollitt and S. Shenoi (Eds.), Springer, Boston, Massachusetts, pp. 245–256, 2005.

[51] M. Pollitt, K. Nance, B. Hay, R. Dodge, P. Craiger, P. Burke, C. Marberry and B. Brubaker, Virtualization and digital forensics: A research and teaching agenda, *Journal of Digital Forensic Practice*, vol. 2(2), pp. 62–73, 2008.

[52] B. Rodriguez and G. Peterson, Detecting steganography using multi-class classification, in *Advances in Digital Forensics III*, P. Craiger and S. Shenoi (Eds.), Springer, Boston, Massachusetts, pp. 193–204, 2007.

[53] B. Rodriguez, G. Peterson and K. Bauer, Fusion of steganalysis systems using Bayesian model averaging, in *Advances in Digital Forensics IV*, I. Ray and S. Shenoi (Eds.), Springer, Boston, Massachusetts, pp. 345–355, 2008.

[54] B. Rodriguez, G. Peterson, K. Bauer and S. Agaian, Steganalysis embedding percentage determination with learning vector quantization, *Proceedings of the IEEE International Conference on Systems Man and Cybernetics*, vol. 3, pp. 1861–1865, 2006.

[55] V. Roussev, Y. Chen, T. Bourg and G. Richard, md5bloom: Forensic file system hashing revisited, *Digital Investigation*, vol. 3(S1), pp. 82–90, 2006.

[56] V. Roussev and G. Richard, Breaking the performance wall: The case for distributed digital forensics, *Proceedings of the Fourth Digital Forensic Research Workshop*, 2004.

[57] V. Roussev, G. Richard and L. Marziale, Multi-resolution similarity hashing, *Digital Investigation*, vol. 4(S1), pp. 105–113, 2007.

[58] V. Roussev, G. Richard and L. Marziale, Class-aware similarity hashing for data classification, in *Advances in Digital Forensics IV*, I. Ray and S. Shenoi (Eds.), Springer, Boston, Massachusetts, pp. 101–113, 2008.

[59] V. Roussev, L. Wang, G. Richard and L. Marziale, A cloud computing platform for large-scale forensic computing, in *Advances in Digital Forensics V*, G. Peterson and S. Shenoi (Eds.), Springer, Heidelberg, Germany, pp. 201–214, 2009.

[60] P. Sanderson, Mass image classification, *Digital Investigation*, vol. 3(4), pp. 190–195, 2006.

[61] A. Savoldi and P. Gubian, Data recovery from Windows CE based handheld devices, in *Advances in Digital Forensics IV*, I. Ray and S. Shenoi (Eds.), Springer, Boston, Massachusetts, pp. 219–230, 2008.

[62] A. Schuster, Searching for processes and threads in Microsoft Windows memory dumps, *Digital Investigation*, vol. 3(S1), pp. 10–16, 2006.

[63] A. Schuster, The impact of Microsoft Windows pool allocation strategies on memory forensics, *Digital Investigation*, vol. 5(S1), pp. 58–64, 2008.

[64] M. Shannon, Forensic relative strength scoring: ASCII and entropy scoring, *International Journal of Digital Evidence*, vol. 2(4), 2004.

[65] M. Sieffert, R. Forbes, C. Green, L. Popyack and T. Blake, Stego intrusion detection system, *Proceedings of the Fourth Digital Forensic Research Workshop*, 2004.

[66] H. Simon, *Administrative Behavior*, Macmillan, New York, 1947.

[67] J. Slay and A. Przibilla, iPod forensics: Forensically sound examination of an Apple iPod, *Proceedings of the Fortieth Hawaii International Conference on System Sciences*, 2007.

[68] J. Solomon, E. Huebner, D. Bem and M. Szezynska, User data persistence in physical memory, *Digital Investigation*, vol. 4(2), pp. 68–72, 2007.

[69] A. Spruill and C. Pavan, Tackling the U3 trend with computer forensics, *Digital Investigation*, vol. 4(1), pp. 7–12, 2007.

[70] C. Swenson, G. Manes and S. Shenoi, Imaging and analysis of GSM SIM cards, in *Advances in Digital Forensics*, M. Pollitt and S. Shenoi (Eds.), Springer, Boston, Massachusetts, pp. 205–216, 2005.

[71] P. Turner, Unification of digital evidence from disparate sources (digital evidence bags), *Proceedings of the Fifth Digital Forensic Research Workshop*, 2005.

[72] P. Turner, Selective and intelligent imaging using digital evidence bags, *Digital Investigation*, vol. 3(S1), pp. 59–64, 2006.

[73] R. van Baar, W. Alink and A. van Ballegooij, Forensic memory analysis: Files mapped in memory, *Digital Investigation*, vol. 5(S1), pp. 52–57, 2008.

[74] C. Vaughan, Xbox security issues and forensic recovery methodology (utilizing Linux), *Digital Investigation*, vol. 1(3), pp. 165–172, 2004.

[75] M. Weier, Hewlett-Packard data warehouse lands in Wal-Mart's shopping cart, *InformationWeek*, August 4, 2007.

[76] S. Willassen, Forensic analysis of mobile phone internal memory, in *Advances in Digital Forensics*, M. Pollitt and S. Shenoi (Eds.), Springer, Boston, Massachusetts, pp. 191–204, 2005.

Chapter 3

TOWARDS A FORMALIZATION OF DIGITAL FORENSICS

Jill Slay, Yi-Chi Lin, Benjamin Turnbull, Jason Beckett and Paul Lin

Abstract   While some individuals have referred to digital forensics as an art, the literature of the discipline suggests a trend toward the formalization of digital forensics as a forensic science. Questions about the quality of digital evidence and forensic soundness continue to be raised by researchers and practitioners in order to ensure the trustworthiness of digital evidence and its value to the courts. This paper reviews the development of digital forensic models, procedures and standards to lay a foundation for the discipline. It also points to new work that provides validation models through a complete mapping of the discipline.

Keywords: Digital forensic models, standards, validation

1. Introduction

Many digital forensic researchers and practitioners have been active in the field for several years. However, it is difficult for a new researcher, particularly one with a narrow technical background, to have a holistic view of the discipline, the tasks involved and the competencies required to carry them out. Similarly, for a new practitioner, the scope and depth of the discipline along with the risks and opportunities are very unclear.

Several issues are in need of discussion. One is that of definition or terminology. What, if any, are the differences between "computer forensics," "forensic computing" and "digital forensics?" In this paper, we use "digital forensics" as an overarching notion that subsumes these terms.

While questions of terminology may be troubling, Pollitt [17] raises the more pressing issue of the quality of digital forensic examinations, reports and testimony in the light of errors in cases brought before U.S. courts over the years. He asks whether different policies, quality man-

G. Peterson and S. Shenoi (Eds.): Advances in Digital Forensics V, IFIP AICT 306, pp. 37–47, 2009.

uals, validated tools, laboratory accreditations and professional certifications would have made a difference in these cases. He calls for practitioners to examine all their methods and to expose them to external
review to ensure that they are trustworthy.

This paper reviews the development of models, procedures and standards underlying digital forensics to provide a foundation for the discipline. The foundation will help ensure the soundness and reliability of
digital forensic processes and the veracity of evidence presented in court.

2.   Digital Forensics

Pollitt [15] provides one of the earliest definitions of digital forensics:

> "[Digital] forensics is the application of science and engineering to the
> legal problem of digital evidence. It is a synthesis of science and law.
> At one extreme is the pure science of ones and zeros. At this level,
> the laws of physics and mathematics rule. At the other extreme, is the
> courtroom."

Pollitt's definition is a foundational one in that it encompasses the digital
forensic process and the possible outcomes. The analysis of computer
systems is a clear goal, but the results must be legally acceptable.

McKemmish [10] describes digital forensics in the following manner:

> "[Digital forensics] is the process of identifying, preserving, analyzing
> and presenting digital evidence in a manner that is legally acceptable."

McKemmish suggests that digital forensics is a multi-disciplinary domain. On the other hand, Palmer [13] defines digital forensics as:

> "The use of scientifically derived and proven methods for the preserva-
> tion, collection, validation, identification, analysis, interpretation, doc-
> umentation and presentation of digital evidence derived from digital
> sources for the purpose of facilitating or furthering the reconstruction
> of events found to be criminal, or helping to anticipate unauthorized
> actions shown to be disruptive to planned operations."

Palmer's definition was developed at the Digital Forensic Research Workshop (DFRWS). In fact, it represented the consensus of the academic researchers in attendance. This definition of digital forensics also implies
the notions of reliability and trustworthiness.

Digital forensics is concerned with investigations into misuse, and its
outcomes must be acceptable in a court of law or arbitration proceedings.
Therefore, there is a heavy reliance on the non-technical areas of the field,
especially given its multidisciplinary nature. This issue is emphasized
by Yasinsac and colleagues [22]:

> "[Digital] forensics is multidisciplinary by nature, due to its foundation
> in two otherwise technologically separate fields (computing and law)."

Several other authors have attempted to define digital forensics. Some explore the scientific and/or legal validity of digital evidence. For example, Bates [1] states:

> "The purpose of forensic investigation is to enable observations and conclusions to be presented in court."

Forte [6] writes:

> "The simple guiding principle universally accepted both in technical and judicial spheres, is that all operations have to be carried out as if everything will one day have to be presented before a judge."

Meyers and Rogers [11] draw attention to the validity of forensic methods and the use of digital evidence in courtroom proceedings:

> "[Digital] forensics is in the early stages of development and as a result, problems are emerging that bring into question the validity of computer forensics usage in the United States federal and state court systems."

Solon and Harper [19] discuss the fragility of digital evidence and the importance of handling evidence properly:

> "Computer-based digital evidence is very fragile; it can be easily altered, damaged or destroyed by improper handling. If the data has not been dealt with correctly, the judge will not allow it to be used in legal proceedings."

Pan and Batten [14] emphasize the use of systematic and sound procedures for evidence extraction:

> "In order to be usable as evidence in a court of law, [information] needs to be captured in a systematic way without altering it in so doing. Thus, the process of identification and handling of the evidence is of prime concern in a forensic investigation."

All these definitions state that the ultimate goal of digital forensics is to provide legitimate and correct digital evidence in a court of law instead of merely examining computer equipment or analyzing digital data. Thus, digital forensics is not a discipline that is focused entirely on technical issues – as some who enter the field from computer science have been taught to believe. Rather, it is a discipline that embraces both computer techniques and legal issues.

3. Digital Forensic Procedures

Operating procedures are an important issue in the field of digital forensics. The quality, validity and credibility of digital evidence are greatly affected by the forensic procedures applied to obtain and analyze evidence.

Reith and colleagues [18] emphasize the benefits of general procedures in digital forensics:

> "This allows a consistent methodology for dealing with past, present or future digital devices in a well-understood and widely accepted manner. For example, this methodology can be applied to a range of digital devices from calculators to desktop computers, or even unrealized digital devices of the future."

General procedures for digital forensics should be flexible rather than being limited to a particular process or system. Reith and colleagues also identify a number of reasons why standard operating procedures (SOPs) are lacking in many operational laboratories. The reasons include the uniqueness of cases, changing technologies and differing legislation. Many of these issues can be addressed by having flexible SOPs that permit changes within a framework but with clear overall outcomes.

3.1   Research-Based General Procedures

Several researchers (see, e.g., [2, 9]) have focused on general digital forensic procedures. These research-based procedures form the foundation for the development of practitioner-based general procedures. Indeed the efforts undertaken in the context of research-based procedures make it much easier to develop practitioner-based general procedures.

Reith and colleagues [18] are certainly not alone in discussing general procedures and methods for partitioning the various stages in a digital forensic investigation. McKemmish [10] lists four major steps in digital forensic investigations: (i) identification of digital evidence; (ii) preservation of digital evidence; (iii) analysis of digital evidence; and (iv) presentation of digital evidence. The generalized procedures described by McKemmish focus on more than just the technical elements; they specifically cover outcomes applicable to judicial personnel (judges, lawyers and juries) who may have limited technical backgrounds.

McKemmish subsequently refined and extended his work with input from other authors, developing the Computer Forensic - Secure, Analyze, Present (CFSAP) model [12]. There are two principal differences between the CFSAP model and McKemmish's earlier work. First, the evidence identification and preservation phases in McKemmish's earlier model are combined to create the secure phase in the CFSAP model. This modification was deemed necessary because the boundaries between the two phases are sometimes not clear.

The second difference is the flowchart used to describe the CFSAP model. This modification gives users a better understanding of the procedures involved in digital forensic investigations. Moreover, a feedback loop permits movement from the analyze phase back to the secure phase to ensure that digital evidence is not overlooked.

Palmer [13] proposed an alternative model at the 2001 Digital Forensic Research Workshop (DFRWS), which we refer to as the DFRWS model. The DFRWS model incorporates six processes, each with its own candidate methods and techniques:

- **Identification:** Event/crime detection, signature resolution, profile detection, anomaly detection, complaint resolution, system monitoring, audit analysis.

- **Preservation:** Case management, imaging, chain of custody, time synchronization.

- **Collection:** Preservation, approved methods, approved software, approved hardware, legal authority, lossless compression, sampling, data reduction, recovery.

- **Examination:** Preservation, traceability, validation, filtering, pattern matching, hidden data discovery, hidden data extraction.

- **Analysis:** Preservation, traceability, statistical methods, protocols, data mining, timeline, link, special.

- **Presentation:** Documentation, expert testimony, clarification, mission impact, recommended countermeasures, statistical interpretation.

Stephenson [20] also describes a model with six phases. His model expands McKemmish's preservation phase into three phases: preservation, collection and examination. Stephenson's phases are listed below. Note that they provide specific information related to preserving digital evidence.

- **Identification:** Determine items, components and data possibly associated with the allegation or incident; employ triage techniques.

- **Preservation:** Ensure evidence integrity or state.

- **Collection:** Extract or harvest individual items or groupings.

- **Examination:** Scrutinize items and their attributes (characteristics).

- **Analysis:** Fuse, correlate and assimilate material to produce reasoned conclusions.

- **Presentation:** Report facts in an organized, clear, concise and objective manner.

In addition to highlighting the unique characteristics of the DFRWS model, Reith and colleagues [18] have proposed a processing model called the Abstract Digital Forensics Model (ADFM). ADFM is an extension of the DFRWS model, but it also draws from other sources such as the FBI crime scene search protocol [21]. According to ADFM, a digital forensic investigation has nine key phases:

- **Identification:** Identify and determine the type of the incident.

- **Preparation:** Organize necessary tools, required techniques and search warrants.

- **Approach Strategy:** Dynamically build an approach to maximize the collection of evidence and minimize victim impact.

- **Preservation:** Protect and maintain the current state of evidence.

- **Collection:** Record the physical crime scene and produce a duplicated image of digital evidence via qualified procedures.

- **Examination:** Perform an advanced search for relevant evidence of the incident.

- **Analysis:** Provide an interpretation of the evidence to construct the investigative hypothesis and to offer conclusions based on the evidence.

- **Presentation:** Provide explanations of conclusions.

- **Returning Evidence:** Ensure that the physical and digital assets are returned to their owners.

Carrier and Spafford [3] have proposed the Integrated Digital Investigation Process (IDIP) model, which is based on theories and techniques derived from physical investigative models. The IDIP model is based on the concept of a "digital crime scene." Instead of treating a computer as a substance that needs to be identified, it is treated as a secondary crime scene and the digital evidence is analyzed to produce similar characteristics as physical evidence.

The IDIP model has five phases: (i) readiness; (ii) deployment; (iii) physical crime scene investigation; (iv) digital crime scene investigation; and (iv) review. The purpose of the digital crime scene investigation phase is to collect and analyze digital evidence left at the physical crime scene. This digital evidence must be connected to the incident under investigation. The IDIP model is further broken down into seventeen sub-phases. Interested readers are referred to [3] for additional details.

3.2   Practitioner-Oriented Operating Procedures

Standard operating procedures (SOPs) are usually the ultimate goal of practitioner-based computer forensic models. Proper SOPs are essential for digital forensic practitioners to perform investigations that ensure the validity, legitimacy and reliability of digital evidence [4, 7–9, 16].

Definitions of SOPs have been discussed for several years. Pollitt [16] notes that standards serve to limit the liability for actions by examiners and their organizations. Lin and colleagues [8] emphasize that law enforcement agencies must define SOPs to enable personnel to conduct searches and process cases in a proper manner. Pollitt [16] examines SOPs from a scientific perspective while Lin and colleagues [8] focus more on the legal consequences of using improper SOPs. Several other authors (e.g., [9, 11]) agree that SOPs have profound significance, but emphasize that they should be flexible enough to accommodate the changing digital forensic environment.

Creating a permanent set of SOPs is infeasible. Pollitt [16] states that standards can impede progress and limit creativity. As new problems and tools become available, new methods for solving forensic problems will be created. Therefore, regular updates to SOPs will be necessary to deal with changing technologies and legal environments. Permanent SOPs will eventually become outdated and useless.

In general, digital forensic practitioners prefer to have a universally accepted set of SOPs. However, Palmer [13] argues that it is difficult to create such a set of SOPs because analytical procedures and protocols are not standardized and practitioners do not use common terminology. Reith and colleagues [18] reinforce this position by pointing out that forensic procedures are neither consistent nor standardized.

4.   Accreditation Standards

Many organizations have sought to maintain high quality in their digital forensic processes by pursuing ISO 17025 laboratory accreditations. This international standard encompasses testing and calibration performed using standard, non-standard and laboratory-developed methods. A laboratory complying with ISO 17025 also meets the quality management system requirements of ISO 9001.

The high workloads and dynamic environments encountered in digital forensic laboratories can make it difficult to meet accreditation requirements. A Scientific Working Group meeting in March 2006 at Australia's National Institute of Forensic Science [5] addressed the principal issues regarding accreditation. While earlier discussions had concentrated on the trustworthiness of digital evidence, the meeting participants strongly believed that the Australian approach should focus on the validation of digital forensic tools and processes.

Given the high cost and time involved in validating tools and the lack of verifiable, repeatable testing protocols, a new sustainable model is required that meets the need for reliable, timely and extensible validation and verification. Also, a new paradigm should be adopted that treats a tool or process independently of the mechanism used to validate it.

If the domain of forensic functions is known and the domain of expected results is known, then the process of validating a tool can be as simple as providing a set of references with known results. When a tool is tested, a set of metrics can also be derived to determine the fundamental scientific measurements of accuracy and precision.

Mapping the digital forensics discipline in terms of discrete functions is the first component in establishing a new paradigm. The individual specification of each identified function provides a measure against which a tool can be validated. This allows a validation and verification regime to be established that meets the requirements of extensibility (i.e., the test regime can be extended when new issues are identified), tool neutrality (i.e., the test regime is independent of the original intention of the tool or the type of tool used), and dynamically reactive (i.e., testing can be conducted quickly and as needed).

The Scientific Working Group agreed to describe the digital forensic component in terms of two testable classes, data preservation and data analysis. These two classes in their broadest sense describe the science in sufficient detail to help produce a model that is useful for accreditation purposes, not only for validation and verification, but also for proficiency testing, training (competency) and procedure development.

Figure 1 presents the validation and verification model. The model covers data preservation and data analysis. Data preservation has four categories while data analysis has eight categories.

A breakdown of the forensic copy category can be used to illustrate the depth of the categorization of functions. Static data is data that remains constant; thus, if static data is preserved by two people one after the other, the result should remain constant. An example is a file copy or a forensic copy (bit stream image) of a hard disk drive. Dynamic data is data that is in a constant state of flux. If dynamic data is preserved once and then preserved again, the results of the second preservation could be different from those of the original preservation. For example, a

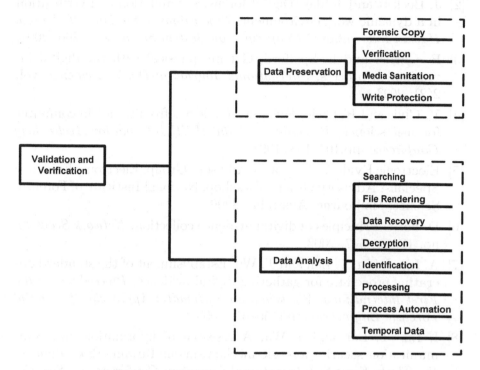

Figure 1. Validation and verification model.

TCP/IP traffic intercept or a memory dump is a snapshot at the instant of collection. A preservation of TCP/IP traffic or computer memory cannot be repeated with consistent results.

5.   Conclusions

Experience has shown that the quality of digital forensic investigations is enhanced by the application of validated procedures and tools by certified professionals in accredited laboratories. Our review of digital forensic models, procedures and standards is intended to provide a foundation for the discipline. We hope that the foundation will help ensure the soundness and reliability of digital forensic processes and the veracity of the evidence presented in court.

References

[1] J. Bates, Fundamentals of computer forensics, *Information Security Technical Report*, vol. 3(4), pp. 75–78, 1998.

[2] J. Beckett and J. Slay, Digital forensics: Validation and verification in a dynamic work environment, *Proceedings of the Fortieth Annual Hawaii International Conference on System Sciences*, p. 266, 2007.

[3] B. Carrier and E. Spafford, Getting physical with the digital investigation process, *International Journal of Digital Evidence*, vol. 2(2), 2003.

[4] V. Civie and R. Civie, Future technologies from trends in computer forensic science, *Proceedings of the IEEE Information Technology Conference*, pp.105–108, 1998.

[5] Electronic Evidence Specialist Advisory Group, Electronic Evidence Specialist Advisory Group Workshop, National Institute of Forensic Science, Melbourne, Australia, 2006.

[6] D. Forte, Principles of digital evidence collection, *Network Security*, no. 12, pp. 6–7, 2003.

[7] A. Lin, I. Lin, T. Lan and T. Wu, Establishment of the standard operating procedure for gathering digital evidence, *Proceedings of the First International Workshop on Systematic Approaches to Digital Forensic Engineering*, pp. 56–65, 2005.

[8] I. Lin, T. Lan and J. Wu, A research of information and communication security forensic mechanisms in Taiwan, *Proceedings of the Thirty-Seventh International Carnahan Conference on Security Technology*, pp. 23–29, 2003.

[9] I. Lin, H. Yang, G. Gu and A. Lin, A study of information and communication security forensic technology capability in Taiwan, *Proceedings of the Thirty-Seventh International Carnahan Conference on Security Technology*, pp. 386–393, 2003.

[10] R. McKemmish, What is forensic computing? *Trends and Issues in Crime and Criminal Justice*, no. 118 (www.aic.gov.au/publications /tandi/ti118.pdf), 2002.

[11] M. Meyers and M. Rogers, Computer forensics: The need for standardization and certification, *International Journal of Digital Evidence*, vol. 3(2), 2004.

[12] G. Mohay, A. Anderson, B. Collie, O. de Vel and R. McKemmish, *Computer and Intrusion Forensics*, Artech House, Norwood, Massachusetts, 2003.

[13] G. Palmer, A road map for digital forensic research, *Proceedings of the 2001 Digital Forensic Research Workshop*, 2001.

[14] L. Pan and L. Batten, Reproducibility of digital evidence in forensic investigations, *Proceedings of the 2005 Digital Forensic Research Workshop*, 2005.

[15] M. Pollitt, Computer forensics: An approach to evidence in cyberspace, *Proceedings of the Eighteenth National Information Systems Security Conference*, pp. 487–491, 1995.

[16] M. Pollitt, Principles, practices and procedures: An approach to standards in computer forensics, *Proceedings of the Second International Conference on Computer Evidence*, pp. 10–15, 1995.

[17] M. Pollitt, Digital orange juice, *Journal of Digital Forensic Practice*, vol. 2(1), pp. 54–56, 2008.

[18] M. Reith, C. Carr and G. Gunsch, An examination of digital forensic models, *International Journal of Digital Evidence*, vol. 1(3), 2002.

[19] M. Solon and P. Harper, Preparing evidence for court, *Digital Investigation*, vol. 1(4), pp. 279–283, 2004.

[20] P. Stephenson, Modeling of post-incident root cause analysis, *International Journal of Digital Evidence*, vol. 2(2), 2003.

[21] K. Waggoner (Ed.), Crime scene search, in *Handbook of Forensic Services*, Federal Bureau of Investigation, Quantico, Virginia, pp. 171–184, 2007.

[22] A. Yasinsac, R. Erbacher, D. Marks, M. Pollitt and P. Sommer, Computer forensics education, *IEEE Security and Privacy*, vol. 1(4), pp. 15–23, 2003.

II

FORENSIC TECHNIQUES

Chapter 4

BULK EMAIL FORENSICS

Fred Cohen

Abstract   Legal matters related to unsolicited commercial email often involve several hundred thousand messages. Manual examination and interpretation methods are unable to deal with such large volumes of evidence. Furthermore, as the actors gain experience, it is increasingly difficult to show evidence of spoliation and detect intentional evidence construction. This paper presents improved automated techniques for bulk email analysis and presentation to aid in evidence interpretation.

Keywords: Unsolicited commercial email, bulk forensic analysis

1.   Introduction

This paper focuses on the examination and interpretation of large email collections as evidence in legal matters. The need for bulk examination methods has become more important because of the high volume of emails involved in unsolicited commercial email (UCE) cases in which one party accuses the other of numerous violations of the law in sending such email.

Current laws typically include statutory damages on the order of $1,000 per email message in cases involving fraudulent email [3, 4, 11]. Some plaintiffs are tempted to acquire and/or produce large volumes of email messages and file suits for millions of dollars. They may configure their environments to accept as many email messages as possible and may involve multiple states in the transmission of email to trigger additional damages on a per state basis. The plaintiffs in some of these cases work together in a loose knit group and use the leverage of high volumes to make the potential risk of litigation very high while driving up defense costs [7]. The plaintiffs acknowledge these techniques and sometimes assert that they are activists seeking to make bulk emailers pay a high price for sending unsolicited email.

G. Peterson and S. Shenoi (Eds.): Advances in Digital Forensics V, IFIP AICT 306, pp. 51–67, 2009.

Defendants in these cases range across a wide variety of companies. Some appear to be criminal enterprises that violate contracts with multiple marketing firms, lease email platforms from criminal groups to send high volumes of email, regularly flout federal and state laws, steal credit card and other related information used in transactions they facilitate, and when sued, shut down and relocate (to Argentina in at least one case). Other defendants are longstanding advertising firms who – almost without exception – seek to follow the laws regarding advertising, including those related to UCE.

From a technical standpoint, bulk email solicitations involve companies that specialize in different facets of the business. Some create and provide advertising copy and images to their clients or place them on web servers, others send emails to large lists of recipients that they maintain in databases, others handle orders and/or fulfillment, yet others process credit cards and other financial instruments. These companies often subcontract with each other, creating a thriving, competitive market in which entities have intellectual property of different types and enter into arrangements with different customers and vendors. The companies often have exclusive arrangements so that an advertisement will only generate leads to the originator. In many cases, competitors use the resources of other companies (e.g., image servers) without permission, or collect contractually exclusive leads from an inserted advertisement and resell them to their customers.

In the case discussed in this paper [10], the plaintiff asserted that 12,576 email messages were sent by the defendant to the plaintiff in violation of statute [3] and requested damages of about $16 million. The case was eventually ruled in favor of the defendant. Our analysis, while covering both sides, ultimately represents the defendant's perspective more than the plaintiff's perspective. For pedagogic reasons, techniques and results associated with other cases are included without distinguishing them.

2.   Challenges

The complexity of Internet business operations complicates the efforts of the plaintiff and defendant. It is often hard to attribute actions to actors, but this is necessary to win a case. Differentiating what came from where, whether images used were actually part of a particular collection, whether a party was making unauthorized use of a competitor's image server, whether emails were in fact from the company whose image server was used, attributing multiple emails to one source when they come from many different addresses and have differing content, and other similar

challenges can be daunting. Even the associations of domain owner-
ship, domain names and IP addresses are often complicated by the large
numbers of domains, addresses and content, the high rate of change of
this information over time, and the lack of timely lookup of relevant information. Furthermore, opponents are not typically cooperative; they
obfuscate whenever feasible; sometimes they refuse to answer questions
or do not provide documents upon request; they do not retain adequate
records or may intentionally destroy records.

Large volumes render the detailed examination of each email sent by
an individual much too time consuming for the legal calendar to sustain.
It is common for a few CD-ROMs of new evidence to be proffered within
a few days of an expert report deadline, or a day or two before a deposition involving the individual identified as knowledgeable about the
content. Evidence also commonly includes content that, upon inspection, leads to additional sources of evidence that have to be identified
and sought. From a tactical standpoint, this evidence is sometimes provided in an obscure form and as a small part of a large collection of
other content, perhaps as a scanned printout of an extraction of a log
file included in tens of thousands of pages of other material.

These and other challenges point to the need for tools that can au-
tomate many aspects of analysis while supporting interpretation by the
expert in a timely and accurate manner. Furthermore, it is important
to be able to apply and modify these tools as new information appears.

3. Tools and Techniques

The most common tools used for analysis are small programs involving
Perl scripts, shell scripts and Unix commands such as `grep` and `awk`.

3.1 Application of Common Tools

Using common tools presents certain tradeoffs. Writing or modifying
scripts on short notice can lead to difficulty in verifying their operation.
Off-by-one errors, misses and makes are commonplace [1]. For example,
if a directory contains a set of files corresponding to what is purported to
be one email per file and the goal is to find the number of files containing
some critical content element, a typical script might be:

```
grep "critical content element" * | wc
```

Two problems with this script are: (i) multiple instances of the string
on different lines in one file cause a miscount of the number of files
containing the content; and (ii) the occurrence of more than one instance
on a single line cause an undercount of the number of instances in the

collection. With thousands of emails, a count of 7,543 that is off by one is hardly substantial, unless the email left out is unique in some manner. But offering the wrong count may produce a challenge from the other side and may degrade the standing of the expert and the quality of the report. Several approaches are available to deal with these sorts of errors. The most important step is to clearly define the objective of the analytical process and to properly report the results.

3.2   Issues of Legal Definition

In one case [10], a key issue was the number of applicable emails. The plaintiff asserted that there were 12,576 "emails," but the evidence provided contained 1,421 "actual emails," i.e., sequences of bytes of the proper format from the proffered file corresponding to what a user would consider an email [8]. The legal definition in this case counted email messages once for each recipient, leading to multiple counts of a single email. Even so, this definition did not clarify how the 1,421 actual email messages became the 12,576 emails asserted by the plaintiff.

3.3   Date and Time

Another issue was the relevant dates for the suit [10]. Because of statutes of limitations, effective dates of laws and legal filing dates, authoritative dates and times of events can be very important. Dates and times in emails depend on several factors. The content of proffered emails may not be trustworthy; dates and times stamped by computers may differ from real-world dates and times; and because time passes as email is in transit, an email sent before a deadline can arrive after it.

An anchor email was used in the case to rehabilitate dates and times [10]. This leveraged the fact that the plaintiff's emails were handled by the vendor Postini, which put date and time stamps on the emails in transit. While the collection of emails may have been forgeries, the assumption that they were not led to the use of the Postini date and time stamps as anchors. Independent contemporaneous emails were used to independently validate Postini date and time stamps. These emails, which had known date and time characteristics, were exchanged between systems under the control of the experts and went through the same Postini servers during the period in question. This reconciliation of date and time information excluded all but 242 of the emails in the case.

There are clearly other date and time issues related to email messages. One of the bases for legal claims stems from damages due to the reduction in available bandwidth, storage, CPU time or other resources. Evidence of damage must be in tangible form and, unless detailed records

Table 1. Extracted email arrival and delay times.

Arrival Time	Delay Time
06/27/02 07:33 AM	+0000-00-00 00:00:02
06/27/02 07:53 AM	+0000-00-00 00:00:06
06/27/02 09:11 AM	+0000-00-00 00:00:04
06/27/02 11:55 AM	−0000-00-00 00:00:03
06/27/02 02:41 PM	+0000-00-01 21:24:25
06/27/02 06:23 PM	+0000-00-01 13:06:42
06/27/02 08:12 PM	+0000-00-01 20:16:02
06/27/02 08:24 PM	+0000-00-01 13:09:01
06/27/02 09:12 PM	+0000-00-02 01:12:32

are kept, this is hard to show. One way to demonstrate damage is to analyze `Received:` headers of emails that show arrival times at servers [6]. The fundamental task is to demonstrate a correlation between these times and email volumes.

The analysis of `Received:` headers is complicated by the use of multiple time zones and time differentials between computer date and time settings. The approach used in [10] was to recast all dates and times in Universal Coordinated Time (UTC) and then examine time differences from hop to hop, where each "hop" corresponded to the `Received:` time stamp of a computer in the processing sequence. This involved: (i) parsing all the `Received:` headers; (ii) normalizing times to UTC; (iii) determining the distance (in hops) from final arrival point for each header; (iv) correlating paths through the email system so that comparable paths are compared with each other and not with other paths; (v) identifying time differentials by hops for common paths as a function of time; and (vi) relating these time differences to email volumes.

Despite the analysis, some puzzling outcomes were encountered. Some emails traveling along certain paths were delayed by days whereas other emails of similar size and content and sent along the same paths either earlier or later arrived within seconds. No crashes or other disruptions during the same time frames occurred to explain the anomalies, and they remain unexplained to this day. An inverse relationship between volumes of emails and delivery times can lead to the conclusion that these emails actually improved system performance. But this is ridiculous because correlation is not causality.

Table 1 presents email arrival times and delays. Each row corresponds to a different email message and the arrival times are sequenced in chronological order. The time delay is the interval between the first

Table 2. Number of emails arriving at different hops by date.

Date	Hop 1	Hop 2	Hop 3	Hop 4	Hop 5
10/01/03	4	4	3	2	0
10/02/03	9	9	9	9	0
10/03/03	8	8	8	8	0
10/04/03	6	6	6	6	0
10/05/03	11	11	10	10	0
10/06/03	11	9	8	7	0
10/07/03	23	20	19	18	1
10/08/03	11	11	11	11	0
10/09/03	12	9	6	6	0

arrival at the plaintiff's servers and the final internal delivery. Note that the emails with delivery times in excess of one day (non-zero `yyyy-mm-dd` value for the delay time) arrived before and were delivered after those processed and delivered within seconds of arrival. The emails were unexceptional in size, makeup and content. This appears to refute claims that emails were delayed by high volume.

Table 2 shows the number of arrivals at different hops in the plaintiff's infrastructure on different days (Hop 1 is the final hop). While some emails could arrive just before midnight and be delivered early the next morning or could be at different distances from their final destinations, in this case, none fit this pattern. All the emails had at least three internal hops before delivery and their times were consistently within a few seconds. Detailed examination showed that some of the excess emails had long delays and others were duplicates generated by the plaintiff.

3.4   Deliverability of Emails

Emails asserted must be "deliverable" in that there must be a user who can actually receive them. Some plaintiffs configure systems to accept any and all SMTP sequences, causing them to receive misdirected emails, emails to nonexistent users or emails to cancelled accounts. This is problematic because it may constitute interception of private communications, which is illegal in some jurisdictions and may be in violation of policies and/or contracts.

Common legal interpretation is that such actions invite the emails and, therefore, cannot be the basis for claims associated with undesired email transmission. SMTP refuses emails to recipients that do not exist without allowing the server to enter a state where data (header or body) can be received. Any receipt for non-existing users may constitute an invitation. In [10], there were 133 invited emails that could not have been delivered, leaving only 109 actual emails to be considered. In other cases, tens of thousands of emails have been similarly excluded, and courts have ruled that the activity was designed to generate law suits and was not the intent of the statures.

Demonstrating this fact involves discovering user identities. This is done from lists of user names in password files, server logs, configuration files associated with remote access servers, and document requests. In [10], RAS server logs, password files and other discovery led to this information, which had to be correlated with the emails in time to determine when the identities were valid.

3.5 Detecting Duplicates and Near-Duplicates

Identifying the cause of duplication is necessary for a plaintiff to assert the authenticity of records. Also, it enables a defendant to assert that records are spoliated. In [10], eleven actual email messages were duplicates that were somehow produced by the plaintiff's processing. In other cases, many thousands of duplicates have been identified.

Duplicates appear in many forms. The most obvious is an exact copy of an email sequence including headers, body and separator; in this case, a byte-by-byte comparison yields an exact match. The analysis can be performed by computing a hash value for every sequence, sorting the values to identify duplicates and verifying the duplicates via byte-by-byte comparison.

In other cases, only parts of email sequences are identical, such as delivery information, message identifiers, Message-ID fields, dates and times, and the rest of the headers and bodies. These duplicates are problematic for the plaintiff because they may indicate evidence spoliation. Examples of observed matches indicative of spoliation include:

- Identical sequences except for the From separator in mbox files [8].

- Identical sequences except that they indicate additional Received: headers.

- Identical sequences except for different date and time stamps on otherwise identical Received: headers.

- Emails with identical From separators but different headers and bodies.

- Emails with identical headers by different content.

- Emails with indicators of cut and paste operations used by web browsers supposedly sent by automated email mechanisms.

- Emails containing content indicative of being processed by receivers such as the systematic addition of content to email bodies of different formats from different sources.

- Emails with identifiers in headers showing different sourcing than other emails supposedly from identical sources.

These and other indicators of spoliation have been seen in actual cases. However, detection is problematic in large volumes of email because human interpretation is often required to make determinations about legitimacy.

The general detection approach is to create matching software that performs imperfect matches of portions of evidence. While matching one item to a large set of other items is straightforward, matching each of n items to each other item requires $O(n^2)$ time. Other methods trade off space for time: using hash values requires $O(n)$ time for hash generation followed by $O(log\ n)$ time for sorting, resulting in $O(n\ log\ n)$ time. This may have to be done for each of a large number of different types of matches. For example, if the removal of each line in the headers is to be considered, this comes to the average length of headers multiplied by the previous time. If altered header lines are to be considered, then this has to be broken down further.

A more comprehensive approach looks at evidence in terms of sequences of words or other symbols starting with length one and going up to some maximum sequence length. Each sequence can then be given a unique number and all components (emails, headers or other subsequences) with equal numbers are determined to match to the specified level of similarity. The analyst then has to interpret the meaning of these matches. Unfortunately, this approach leads to very large numbers of matches, and the analyst must again find a way to explore only subsets of the matches in order to keep time and costs down.

One of the most effective approaches is for a human to perform rapid comparisons of sequences of components. Similar components can then be compared in more detail and any obvious mismatches excluded.

3.6   Grouping Extracts for Comparative Analysis

Another approach for detecting anomalies in analysis and interpretation is to create an error model and look for identified error types. One class of error types and method for grouping emails is by the structure of headers. While header lines are largely unstructured, they normally begin with a sequence of characters followed by a colon (:) and continue on lines that start with whitespace [6]. A simple parser can add the lines starting with whitespace to the previous lines starting with a header identifier to allow line-by-line parsing, which makes parsing and analysis easier to program.

The extraction of email sequences from an `mbox` file produces a set of sequences ("extracts"). Extraction of header lines from extracts and identification by extract, line in extract and header identifier allows a wide range of analytical techniques to be applied with relative ease. Using disjunctions, conjunctions and other similar operations allows easy analysis such as the detection of all emails with no `cc:` field containing a particular domain name. The analyst can then use these operations on emails to find similarities and differences relevant to the case at hand.

In [10], it was important to identify emails containing particular IP address ranges in the `Received:` headers as recorded by the plaintiff's computers. Extracting this data is non-trivial because `Received:` lines have a non-standard format. However, once a parser is designed for the particular header lines of the mail transfer agent (MTA) software in use, the IP addresses can be associated with extracts, lines in extracts and the `by` portion of `Received:` headers to include only the desired extracts. Claims regarding these extracts could then be analyzed by customizing other analysis program snippets for the specific claims.

Even in cases involving more than 100,000 emails, the separation of email extracts by headers and the analysis of each header can be completed in a day or two. This tends to yield a great deal of information about emails. Simple sorting of headers rapidly yields information about similarities and differences. Types of information that can be detected include:

- Headers that are misspelled or otherwise differ from normal expectations.

- Associations of emails to other emails based on unique header fields or other header content.

- Sequencing information about the infrastructures involved in email transport.

- Details of the protocols, MTAs, hardware and software involved.

- Attribution information associated with unique identifiers.

- Groups of emails apparently sent from, through or by the same or similar MTAs, systems and mechanisms.

The following is a tree depiction of an email handling process (with details intentionally obfuscated). Such a tree can be generated by analyzing `Received:` headers and may include a variety of subfields depending on analyst needs.

```
        0 325802 B.net
          1 325090 mail.R.com
            2 325090 mail.R.com
              3 215585 mail.H.com
                4 232   mail.R.com
                4 24    other.H.com
              3 109301 other.H.com
                4 5     mail.R.com ...
```

In the tree above, almost all the emails (325,090 of 325,802) arriving at B.net arrived through mail.R.com, all of which came again through mail.R.com, most of which came through mail.H.com with most of the remainder coming through other.H.com. In this case, a close relationship exists between B.net, R.com and H.com. Interestingly, some emails originally arriving at mail.R.com go through mail.H.com, and back to mail.R.com before being delivered to B.net. Depending on where the various servers are located, this looping between providers may be evidence of intentional forwarding of emails through multiple jurisdictions to add damages to the legal action.

Given the definitions used by the plaintiff's expert in [10], the total number of emails that could be the issue came to only 175 out of the original asserted claim of 12,576. These included 98 actual emails combined with the 34 unique active recipient addresses that were potential recipients of the actual emails. This analysis alone reduced the potential damages from more than $10 million to less than $200,000. But, as we discuss below, this was not the end of the issue.

3.7   Signups, Invitations and Other Causes

Another substantial limit on UCE cases is that laws tend to exonerate defendants who fulfill user requests or email users with a pre-existing relationship. A user who requests information on a web site may intentionally or inadvertently agree to terms and conditions granting the right to send or cause to be sent email that would otherwise be categorized as UCE.

When large volumes of emails are involved, it may become problematic for the plaintiff to prove that the emails were not solicited or that the addressees requested cessation of the emails. The plaintiff presumably has to show that the emails in question were unsolicited, which requires the presentation of legal documents signed by the individual recipients or proffering evidence associated with things like customer complaints to make the case. For hundreds of thousands of emails sent to hundreds or thousands of recipients, this would involve an enormous quantity of paperwork and would cause disturbance to numerous customers.

In practice, high volume UCE cases usually involve a small number of individuals who act on their own or assert their role as email service providers to sue defendants repeatedly. They try various methods to enable them to assert large numbers of emails, such as taking over the accounts of previous users, allowing emails directed to other recipients to be sent to them, making copies of emails sent to users and resending the duplicates to themselves, and forwarding emails through multiple jurisdictions to create additional penalties. These tricks may work in cases that are poorly defended, and some plaintiffs have won high-valued default judgments against defendants. One such defendant lost suits totaling more than one billion dollars; but the owners closed their business and left the country years ago (after apparently committing fraudulent activities). It seems unlikely that the judgments will ever yield real compensation. When plaintiffs create the conditions associated with high volumes of UCE, defendants are often able to show that the emails were invited. For example, when email identified as "spam" is sent, resent or forwarded, this may constitute an invitation and the forwarder may be liable for the violation.

From a technical standpoint, showing that protocols would not or could not have sent the emails unless the plaintiff acted to enable them may be adequate. Plaintiffs have an obligation to mitigate damages as well. For example, configuring email servers to allow all emails (not just to known users) to be sent or forwarding to a "dummy" account are clear indications of inviting emails. Showing the technical basis for such claims typically means examining log files, configurations and ancillary information, testing configurations in reconstructions with data from the case, and showing that the configurations produce the results at issue. In some cases, only screenshot images are available that are purported to depict the configuration of a product. Lacking version information and other relevant material, the analyst must ultimately make assumptions and draw conclusions based on the assumptions. But the evidence can often help. For example, in one instance, a configuration screen clearly indicated that the MTA was configured to forward emails identified as "spam" to a third party under a different email address. This party intentionally and knowingly sent the emails in question to the plaintiff, making him potentially liable.

When emails to users are copied, potential liability arises based on contracts with users or privacy regulations. To the extent that plaintiffs do not retain such information or fail to produce it, preservation has been held to be required as soon as a plaintiff is aware of the potential for legal action [2]. While there is a duty for UCE mailers in the United States to retain information on removal requests and not send additional emails, signups are typically held as proprietary information by vendors involved in different aspects of the business. For example, an advertiser almost certainly would not have information about the individuals who receive its advertisements and can only process removals by providing them to the solicitor. Contracts between advertisers and UCE mailers typically provide for the timely removal of users from mailing lists and include requirements for following applicable laws and regulations. This presumably limits the liability of advertisers, but laws vary on how this liability may apply to entities who order the insertion of advertisements.

Other causes may be asserted for individual emails. In one instance, an email was shown to have been received a second time six months after it was originally delivered. This makes the case for examining system logs and information on system failures, crashes, reboots, break-ins and other events that might cause delays. The example mentioned above appears to be the result of a restoration from old backups after a crash.

3.8   Compliance with Internet RFCs

In many cases involving high volumes of UCE, plaintiffs have claimed that the headers were false or misleading based on their compliance or non-compliance with Internet RFCs [6]. As of this time, courts have not ruled that RFCs constitute legal contracts or are enforceable in a legal sense. Nevertheless, claims are made with regard to RFCs in many cases and expert witnesses are called upon to testify with regard to RFCs, their interpretations and the extent to which they may have been violated.

- **HELO Lines:** One of the most common assertions made by plaintiffs in these cases is that a "HELO" indicates a fraudulent source. The HELO exchange is used in the initiation of an SMTP (RFC 821 [6]) exchange in which the sending computer is supposed to send HELO followed by a string. The HELO information is sometimes recorded in a `Received` line associated with that hop in the delivery process. RFC 2821, the updated version of SMTP, uses "EHLO" instead of HELO to initiate its processing, indicating to the receiving server that the RFC 2821 protocol applies. RFC 2821 indicates that in cases when the HELO protocol is used, RFC 821 must be used to process email.

  Most of the emails we have seen in bulk email cases conform with RFC 821 instead of RFC 2821. RFC 821 specifies domain names such as `localhost`, but it does not assert any requirement of authenticity. Email recipients never see the HELO lines sent to SMTP servers unless they examine log files associated with the email. Moreover, the recording of HELO information is neither mandatory nor is it intended for users. The normal presentation of an email does not include the area that contains HELO information. Many commonly used email clients have versions that send the name of the receiving (instead of the sending) computer in the HELO line apparently because the authors misread the RFCs.

Filtering based on the HELO information is sometimes used to prevent emails from known undesirable source domains. Some MTAs check the IP address against the domain name using a DNS number-to-name lookup and place a warning in the header to notify spam filters of a mismatch. However it is extremely common for DNS names and IP addresses to not match in a number-to-name lookup for several reasons, including when (i) large numbers of domain names are associated with a single IP address; (ii) proxy servers are used for delivering email; (iii) email delivery services are used for delivering email; (iv) servers are named incorrectly during configuration; (v) default server names are not updated during a configuration; and (vi) emails are sent from mobile locations (e.g., coffee shop, bookstore or hotel room). Dynamic DNS introduces additional complications and multiple answers to name-to-address lookups are not compensated for in many reverse lookup approaches.

- **False Sender Identities:** Another common claim by plaintiffs is that the use of a fictitious name or email address in a sender identity (e.g., `From:` line) is deceptive. Some plaintiffs have claimed that the use of an email address not containing the name of the sender is fraudulent because it misleads the recipient into believing that the sender is someone he is not. While this might appear to be a cogent argument, Internet systems often use fictitious names and pseudonyms, including names like `accounting` in RFC 821 and a wide variety of other sender names in emails from almost any company that can be identified.

In most of the cases where this claim has been made, the plaintiff also uses false names as do the plaintiff's providers and customers, making the claim that much more problematic. However, the issue is not all that clear in law. There is a real possibility that some court will eventually rule differently, making pseudonyms and anonymized names problematic as well.

Experts called to testify about Internet conventions and other common usage may examine the use of naming by the plaintiff and defendant, their ISPs, other providers, supply chain entities and government agencies, including the court of competent jurisdiction. We have found that it is a cogent argument to demonstrate that the court making the ruling does the very thing the plaintiff claims to be fraudulent. While some may decide to give an opinion about a fictitious name being misleading, this is problematic. Unless the digital forensic expert is also an expert in linguistics, he risks having his credibility destroyed along with the rest of his testimony.

Forged sender identities may be identified by an expert so long as there is a basis for showing that the user identity was used without the permission of the real person. The potential for forgery already exists, for example, when senders claim to be the recipient or use the identity of a different individual. A recent case [5] involved claims of more than 2,000 Usenet postings using a U.S. Chess Federation board member's name to discredit him and gain his board position. Credibly tracing these to the sources then becomes the issue.

- **False Received Headers:** Emails are sometimes sent with forged `Received:` headers to mislead recipients who attempt to trace email. These are problematic in individual email cases when forgers use realistic sequences. But such forgeries are not as trivial as they may appear, especially in volume cases because of timing and consistency problems with forgeries and the use of legal means to obtain records from other sites. Usually such forgeries involve a common intermediary associated with many other reception sequences, which are easily detected when presented in a tree format as shown in Section 3.6.

3.9   Inconsistencies

The examination of subject lines for deceptive content typically requires a linguistics expert. However, technical analysis has shown inconsistency in claims in high volume email cases. Claims typically require the explicit identification of specific statutory violations associated with each asserted email. Since many `Subject:` lines may be identical or nearly identical, analysis for consistency may reveal weaknesses in the plaintiff's claims. In a recent case involving more than 10,000 emails, approximately 30% of the claims about `Subject:` headers were inconsistently made, and the plaintiff lost a summary judgment. Inconsistencies in claims also goes to other issues and should be examined in high volume cases using automated techniques.

3.10   Assessment of Damages

Experts may also be asked to assess damages under trespass laws. Damages in such cases involve physical damage, deprivation, conversion or lost value or rights. In high volume email cases, only deprivation typically applies, and only to the extent that the plaintiff can prove quantified, time-framed, tangible, unmitigatable damage caused by uninvited messages by the defendant [9]. To date, plaintiffs in high volume email cases have largely failed to produce such proof.

3.11   Tracing Emails

To demonstrate causality, it is often necessary to trace emails to their origins. There are three common approaches for determining causality. One is to go from the destination back to the source step by step using subpoenas and gather evidence along the way. This approach has proved to be successful when applied properly.

The second approach takes a shortcut to the origin. In [10], the plaintiff responded deceptively to a UCE by providing a false lead to the seller. When the seller responded to the lead, plaintiff accused the seller of causing UCE to be sent and used this action to conduct a trace from the seller onward. This strategy might have worked if the plaintiff had not lied about the response, which brought up the counterclaim of unclean hands and whether the forward trace had yielded only a single sending chain. In [10], this evidence helped clear the defendant because, as it turned out, a fraudulent intermediary had violated exclusive lead generation contracts by selling the generated leads to many advertising agencies and was, thus, not acting on behalf of the defendant. Problems with this approach are (i) it may not produce a unique sender because of lead sharing; (ii) the entity sending the advertisement may not be the entity who "benefits" from it; (iii) care must be taken to ensure that the process is properly recorded; and (iv) just because one email produces this behavior does not mean that others will produce the same result.

The third approach is to use information in the bodies of emails. This typically involves the assertion that a URL contained within an email is used by a defendant in their business to track or display advertisements. If this is relied upon by the defendant, then the theory is that it should be an adequate record to show that the defendant caused the email to be sent. Problems with this approach include (i) competitors can and regularly do use "image servers" of others in their businesses so that other companies pay for the space, artwork and bandwidth while they gain the financial advantages; (ii) a malicious actor could provide the information for the purpose of damaging the defendant's reputation;

and (iii) someone else could use the URLs for any purpose, including for falsifying the records to create a legal action. Other sorts of commonalities have similar problems, but some success may be gained by using this information in conjunction with the first approach.

4.   Conclusions

Making a case against bulk email senders involves most of the same elements one would use in any legal case involving digital evidence. However, challenges to digital evidence in the larger sense [1] must be met in order to make a case against a competent defense. Key factors that differentiate bulk email cases from other matters are: (i) the evidence must be explored using automation and any automated techniques used must meet legal standards; (ii) contemporaneous records should be properly identified, collected and preserved to obtain the evidence necessary to prove the case; and (iii) increased care should be taken because small mistakes tend to get amplified by volume. Poorly constructed cases, exaggerated claims, spoliated evidence and large volumes of invited emails are likely to be detected by a competent defense, especially in cases involving large monetary claims. In the case [10] discussed in this paper, the defendant won a summary judgment. While digital evidence played a substantial role in the decision, as always, the evidence and analysis are applicable in the context of the specific case. Nevertheless, the techniques may be applied to a variety of bulk email cases.

References

[1] F. Cohen, *Challenges to Digital Forensic Evidence*, ASP Press, Livermore, California, 2008.

[2] C. Crowley and S. Harris (Eds.), The Sedona Conference Glossary: E-Discovery and Digital Information Management, The Sedona Conference, Sedona, Arizona, 2007.

[3] Government of California, Article 18: Restrictions on Unsolicited Commercial E-Mail Advertisers, *West's Annotated California Codes (Business and Professions Code)*, §17500 to §18999.99, pp. 101–117, 2008.

[4] Government of Maryland, Definitions, Commercial Electronic Mail (Subtitle 30), *Michie's Annotated Code of the Public Laws of Maryland (Commercial Law)*, pp. 476–477, 2005.

[5] D. McClain, Member of U.S. Chess Federation's board is asked to resign in dispute over an election, *New York Times*, January 15, 2008.

[6] J. Postel, RFC 821: Simple Mail Transfer Protocol, Information Sciences Institute, University of Southern California, Marina del Rey, California (tools.ietf.org/html/rfc821), 1982.

[7] SpamLinks, Anti-spam laws (spamlinks.net/legal-laws.htm).

[8] Sun Microsystems, `mbox`, Manual pages from `/var/qmail` (version 1.01), Santa Clara, California (www.qmail.org/qmail-manual-html /man5/mbox.html).

[9] Supreme Court of California, Intel Corporation v. Hamidi, *West's Pacific Reporter (Third Series)*, vol. 71, pp. 296–332, 2003.

[10] U.S. District Court (Northern District of California), ASIS Internet Services v. Optin Global, Inc., Case No. C-05-5124 JCS, December 17, 2008.

[11] U.S. Government, Controlling the Assault of Non-Solicited Pornography and Marketing Act, Public Law 108–187, 108th Congress, *United States Statutes at Large*, vol. 117(3), pp. 2699–2719, 2004.

Chapter 5

ANALYZING THE IMPACT
OF A VIRTUAL MACHINE
ON A HOST MACHINE

Greg Dorn, Chris Marberry, Scott Conrad and Philip Craiger

Abstract   As virtualization becomes more prevalent in the enterprise and in personal computing, there is a great need to understand the technology as well as its ramifications for recovering digital evidence. This paper focuses on trace evidence related to the installation and execution of virtual machines (VMs) on a host machine. It provides useful information regarding the types and locations of files installed by VM applications, the processes created by running VMs and the structure and identity of VMs, ancillary files and associated artifacts.

Keywords: Virtualization, virtual machine, VMware, Parallels

1.   Introduction

The Sarbanes-Oxley Act of 2002, which was passed as a result of major accounting scandals, requires corporations to have the digital forensic capability to identify and recover information and documentation related to their business operations. As virtualization becomes more prevalent, there is a great need to understand the technology and its impact on evidence recovery.

This paper investigates several key aspects related to the behavior of virtual machines (VMs) and the impact of VMs on host systems. The structure of VMs is explored by experimenting with two popular virtualization systems, VMware [14] and Parallels [11]. The associated VMs are analyzed to identify the artifacts produced during VM creation and deletion and the impact of their processes on the host machine. Also, VM behavior is investigated by deleting individual files from within a VM and observing the results. Commercial off-the-shelf forensic software

G. Peterson and S. Shenoi (Eds.): Advances in Digital Forensics V, IFIP AICT 306, pp. 69–81, 2009.

systems, including EnCase [6] and Forensics Toolkit [1], are used to locate artifacts that remain on the host system during and after the use of virtualization software and to determine if the artifacts are recognizable as belonging to a VM or the host machine.

2.   Testing Methodology

Our method for researching and testing the behavior of virtualization software and VMs involved four steps: (i) virtualization software was installed on a new desktop computer and used to create several VMs; (ii) test files were placed on the VMs, executed and then deleted, and the changes in the VMs and host system were recorded; (iii) the hard drive on the host machine was imaged; and (iv) the image was analyzed using forensic software.

2.1   Host Machine Configuration

A new Dell Optiplex 755 machine (referred to as the host machine) was used for the tests. The drive was wiped using a Knoppix Live CD version 5.1.1 [9] and the command dd if=/dev/zero of=/dev/sda.

Next, Windows XP SP3 was installed, followed by VMware Workstation 6.0.3 [15] and Parallels Workstation 2.2 [12].

2.2   Virtual Machine Configuration

VMware Workstation was used to create two VMs: Windows XP SP3 and Ubuntu 8.04 Desktop [3]. The VMs were created with the default settings within VMware using the "Typical" option. The guest operating system type was chosen from the drop down list, the VM was assigned the default name provided by VMware, bridged networking was assigned as the network connection, and the size of the virtual hard disk was set to the default 8.0 GB.

The Parallels Workstation was also used to create two VMs: Windows XP SP3 and Ubuntu 8.04 Desktop. The VMs were created with the default settings within Parallels using the "Create a Typical VM" option. The guest operating system was chosen using the appropriate option, the VM name set to the default provided by Parallels, a new folder was created by Parallels for the newly created VM files, a 32 GB virtual hard disk was created and bridged networking was used. The guest operating systems were installed using installation CDs or DVDs for each VM in an identical fashion to the installation of the operating system on a physical machine, where the physical machine boots from the installation media and the operating system is installed.

Several types of files were used to test VM behavior: .mp3, .jpg, .htm, .txt, .xls and .pdf. The files were copied to the host machine and to all four VMs via a network shared drive and placed in folders labeled Test Files. The test files for all the Windows machines (including the host machine) were placed on the Desktop and in the My Documents and WINDOWS folders. For the Ubuntu VMs, the test files were placed on the Desktop and in the Home folder. The test files were given different names to identify them by their type, operating system and location. Next, one file was deleted from each location on all the VMs. Then, one VM was deleted from each type of virtualization software, Ubuntu from VMware and Windows XP from Parallels. This was done in order to compare and contrast the behavior of the two types of operating systems across the two virtualization software systems.

2.3   Analysis Machine Configuration

An established Dell Optiplex 745 machine with Microsoft Windows XP SP2 was used as the analysis machine. This machine contained En-Case versions 5.5.11.0 and 6.10.0.25, Forensics Toolkit 1.70.1 and Xways Forensics 14.2 [19]. The machine also contained WinHex [18], Hex Editor Neo [8] and TextPad [7] for viewing and altering files.

2.4   Host Machine Imaging

The hard drive on the host machine was imaged immediately after every major change to the file system. The drive was imaged after the initial configuration was complete, including the deletion of the specified VMs and test files. A second image was created after clearing the Recycle Bin/Trash on the host machine and all the VMs. A third hard drive image was created after the Disk Cleanup tool was used on the virtual drive of the VMware XP VM; the sudo apt-get clean command was executed on the Parallels Ubuntu VM; and the Disk Cleanup and defragmentation tools were used on the host machine.

A special protocol was used for imaging drives. First, the host machine was shut down and its hard drive removed. Next, the removed hard drive was connected to the analysis machine via an UltraBlock SATA Bridge Write Blocker [5]. The image was then acquired using EnCase and saved to a different case file labeled according to the specific state of the drive at the time it was imaged (e.g., Case 10.1 Deleted Files Image).

3.   Virtual Machine Structure

The structure of a VM depends on the type of virtualization software used to create it. Specific artifacts and evidence created by virtualization

Table 1. Virtual machine file types.

VMware	
.vmdk	Virtual hard disk file
.vmx	Virtual machine configuration file
.vmxf	Supplemental configuration file
.nvram	BIOS for virtual machine
.vmsd	Dictionary for snapshots
.log	Virtual machine activity log
Parallels	
.hdd	Virtual hard disk file
.pvs	Virtual machine configuration file

software and VMs include (but are not limited to) registry entries, VM files, processes and virtualized hardware. This information pertains to the standard objects related to VMs and virtualization software and is not exhaustive.

3.1   File System

VMware and Parallels create VMs in a similar manner. A main folder is created for the type of virtualization software, a subfolder for each VM is created in the main folder, and the files comprising the VM are located in that subfolder.

Each folder contains a subfolder for the VM, which is created using a naming scheme to reflect the operating system used by the VM (the default naming scheme). The individual VM folders contain a number of files that comprise the VM. VMware creates a default set of six files [16] while Parallels creates two files (Table 1). Of these files, the most important are the virtual hard disk files (.vmdk and .hdd) and the configuration files (.vmx and .pvs).

3.2   Registry

The host machine on which VMs execute contains several registry entries related to virtualization. These entries, shown in Figure 1, define the types of virtualization software and/or the files that relate to the virtualization software used to create the VMs and the VM configuration files [10].

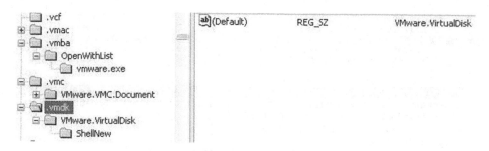

Figure 1. HKEY_CLASSES_ROOT registry entry.

The key shown in Figure 1 was found using the "Find" option in the registry using the value "vmware." Additional searches for the VM files provided the locations of their keys.

3.3 Virtualization Software

One of the most obvious artifacts of virtualization on a host machine is the presence of virtualization software. The two most popular systems of this genre are VMware and Parallels. Open source virtualization software systems include Qemu [2], Xen [4] and VirtualBox [13]; however, these systems are not considered in this work.

Figure 2. Program Files entries.

Most types of software installed on a machine can be identified based on the naming conventions used for their file entries in the host machine file system; these file entries are generally located in the Program Files folder of the host system (Figure 2). Beyond the standard file system entries, additional evidence of virtualization software may persist in the WINDOWS Prefetch folder and the WINDOWS Temp folder. These entries can remain on the host machine even after VMs and the virtualization software are deleted.

3.4 Virtualized Hardware

VMs, like physical machines, rely on hardware to facilitate network connections and to read/write different types of media. To that end, virtualized hardware is created for each VM type (VMware and Parallels).

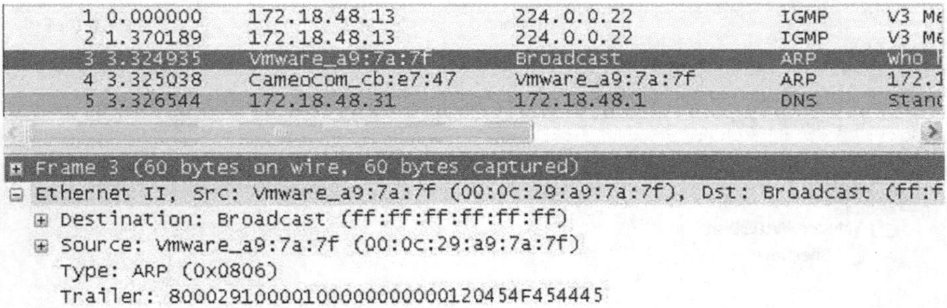

1 0.000000	172.18.48.13	224.0.0.22	IGMP	V3 Me
2 1.370189	172.18.48.13	224.0.0.22	IGMP	V3 Me
3 3.324935	Vmware_a9:7a:7f	Broadcast	ARP	who i
4 3.325038	CameoCom_cb:e7:47	Vmware_a9:7a:7f	ARP	172.1
5 3.326544	172.18.48.31	172.18.48.1	DNS	Stanc

```
⊞ Frame 3 (60 bytes on wire, 60 bytes captured)
⊟ Ethernet II, Src: Vmware_a9:7a:7f (00:0c:29:a9:7a:7f), Dst: Broadcast (ff:f
  ⊞ Destination: Broadcast (ff:ff:ff:ff:ff:ff)
  ⊞ Source: Vmware_a9:7a:7f (00:0c:29:a9:7a:7f)
    Type: ARP (0x0806)
    Trailer: 8000291000001000000000000120454F454445
```

Figure 3.   Wireshark capture of VM traffic.

Virtualized hardware may include a virtual hard drive, virtual display, virtual network device/adapter, virtual USB interface/manager, virtual SCSI/RAID controller and mouse/pointing device.

Evidence of the use of virtualized hardware can be found as well. The virtualized network interface card for a VM identifies itself as a VMware product during a Wireshark [17] capture. Note that the MAC address of the card is also recorded (Figure 3). VMware signatures are indeed pervasive on all types of virtualization hardware.

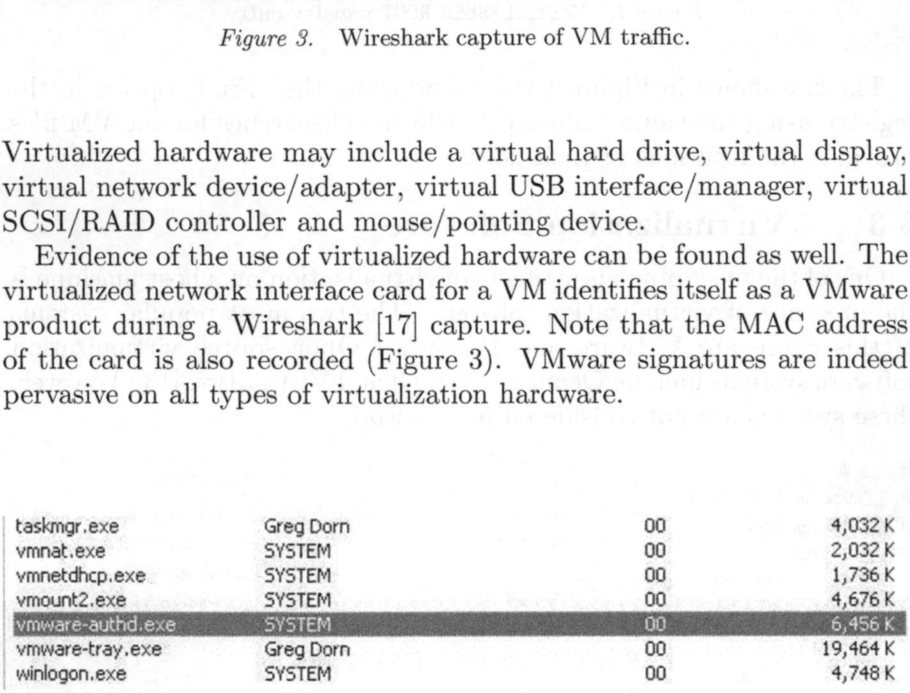

taskmgr.exe	Greg Dorn	00	4,032 K
vmnat.exe	SYSTEM	00	2,032 K
vmnetdhcp.exe	SYSTEM	00	1,736 K
vmount2.exe	SYSTEM	00	4,676 K
vmware-authd.exe	SYSTEM	00	6,456 K
vmware-tray.exe	Greg Dorn	00	19,464 K
winlogon.exe	SYSTEM	00	4,748 K

Figure 4.   Process entries.

3.5   Processes

Additional traces of virtualization software can be located within running processes and the startup configuration of a host machine. Figure 4 shows how these areas within a host machine list the virtualization software that is running on the machine.

VMware and Parallels maintain process items on a host machine whether or not a VM is running. When one these programs is running, the number of processes associated with it increases as when any new program or application executes on a machine.

Table 2. Suspended state and snapshot files.

VMware	
`.lck`	Lock file created when VM is running
`.vmss`	Suspended state file
`-snapshot.vmsn`	Snapshot file for VM reversion
`-snapshot.vmem`	Snapshot of VM memory
Parallels	
`.sav`	Suspended state file

4. Suspended State and Snapshot Artifacts

Beyond the standard set of files created with each new VM, several types of files are created when a VM is suspended or its current state is saved as a backup (snapshot). A key feature related to the suspended state and snapshot files is the listing of configuration settings that detail VM attributes, i.e., the operating system installed on the VM, virtualized hardware attached to the VM, and the paths to any ISOs that were used and hard disk files that were accessed. Table 2 provides a listing of these files.

An examination of the `.vmss` and `-snapshot.vmsn` files indicates that they contain information that is nearly identical to the configuration (`.vmx`) file of a VM, i.e., they contain VM settings. These files can be an important source of information when analyzing a drive from the host machine or VM with regard to the existence of a VM or the behavior of a running VM.

The suspended state files (`.vmss` and `.sav`) are not retained when a VM is restarted. The files are deleted in the same manner as the entire VM – they are not sent to the Recycle Bin and are simply flagged as being available for overwriting. These files can be recovered using the methods described below.

Snapshot files are maintained in the VM folder until they are either deleted individually or deleted as part of the VM. In both cases, the files are sent to the Recycle Bin or Trash. As with the suspended state files and other files associated with VMs, these files are recoverable from the host machine hard drive.

A snapshot function is not available for Parallels Workstation for Windows. However, Parallels for Macintosh does support a snapshot function and creates four different files (`sav`, `.pvc`, `.mem` and `.png`). These

files are similar to their VMware counterparts in that they contain data
about the current state of the VM.

5.   Deleting and Identifying Virtual Machines

A VM can be deleted by deleting it from the virtualization software
or by deleting from the host machine the individual files and folders
that comprise the VM. Deleting a VM from a host machine using the
virtualization software simply places the files back in unallocated space.
The files are not put into the Recycle Bin (like most other files) because
they are typically larger than what the Recycle Bin would allow. Some
files and folders from the host machine could be put in the Recycle Bin
if they are small enough; however, the hard disk file is typically too large
and is, therefore, placed in unallocated space. These files are intact (and
recoverable) until they are overwritten.

EnCase was used to view and analyze the host machine hard disk
images. The contents of the hard disks were viewable and searchable
and the locations for VMware VMs were explored. The two VMs created
with VMware were visible, the Windows XP VM was intact and the
Ubuntu VM was flagged as being deleted. A search of the VM parent
folders revealed files that comprised a VM created with VMware; the
files were intact and were viewable with EnCase.

5.1   Identifying Deleted Virtual Machines

The hard disk file associated with a VM is important because it contains all the information about the VM. The principal difference between
the structure of the VM hard disk file and that of the host machine hard
disk is the information written at the beginning of the disk specific to
the types of headers and other configuration data needed by the VM.
Once the sectors containing the operating system are reached, the VM
and the host machine disks appear identical. We discovered that the
VMware virtual hard disk files begin with the header `KDMV` and Parallels
hard disk files begin with the header `WithoutFreeSpace`. Using these
values, hard disk files can be searched for in the unallocated space on
the host machine drive.

5.2   Recovering Deleted Virtual Machines

During the course of an investigation, it may be necessary to extract
a deleted VM from the unallocated space of the hard disk. Searches
of the unallocated space using known headers yielded the locations of
deleted VMs. Once the hard disk file is located, the file can be treated
as an actual hard disk with respect to locating the Master Boot Record

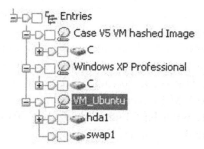

Figure 5.   EnCase view of added VM hard disk file.

within the file. Our research showed that the partition table was typically located at the relative offset 001101ca within the hard disk file for VMware and at the relative offset 003f81da for Parallels. This information allowed us to locate the type and size of the partitions for extraction purposes. VMware typically breaks the virtual hard disk into 2 GB portions until the virtual hard disk size specified by the user is reached; this enables a size value to be used when attempting to extract a VM created by VMware. Once the size of the hard disk file is known, EnCase can be used to extract the file and save it to the analysis machine.

The virtual disk of a VM can be exported to the analysis machine using EnCase and then added back to an open case in EnCase as a new device, rendering the single .vmdk file as if it were another complete hard drive containing an operating system. An example is shown in Figure 5.

At this point, the VM hard disk can be searched for relevant information or files as in any forensic investigation. While other files are necessary to run the VM in the virtualization software, the hard disk file is the most important aspect of a VM.

6.   Deleted Files on a Virtual Machine

It is important to understand how files are treated within a VM. The behavior of a VM hard disk file is identical to that of a physical hard drive on a physical machine. The operating system is installed in the same manner on both types of machines, ensuring that applications and files are handled in a similar manner.

We created a set of test files to identify what happens when a file in an VM is deleted and then the VM itself is deleted. A file deleted in a VM is treated in the same way as one deleted in a physical machine: the file is moved to the Recycle Bin or Trash, where the first bit of the file is flagged to indicate that it can be overwritten as necessary. Recovering a deleted file from a VM is similar to recovering one from a physical machine. Depending on the operating system, this can range from simply restoring the file from the Recycle Bin or Trash to using a third-party application to restore the file.

In the case of a deleted VM, it is necessary to locate and then extract the VM hard disk file in the same manner as a physical drive; this provides access to all the data (intact and deleted) that resides on it. Further analysis of the VM files can be performed by recovering the hard disk file and viewing it with EnCase as described above or by recreating the entire VM as described below.

7.   Recreating a Virtual Machine

Standard forensic analysis focuses on the hard drive of a target machine and possibly on the additional hardware installed on the machine. Virtualization changes this perspective because of the introduction of virtual software and hardware. It is possible – and sometimes necessary – to recreate intact and deleted VMs on a host machine.

In our experiments, the VMware Workstation Windows VM was found to be intact in the file system of the host machine. Once the appropriate files were located for the VM, the `Copy/Unerase` function was used to copy them to a folder called RecoveredWindows on the analysis machine. As each file was copied, it was renamed to `recovered(VMware file type)` (e.g., `recoveredWindows XP Professional.vmdk`). Then, the newly created folder was opened and the configuration (`.vmx`) file was opened, which automatically opened VMware Workstation. The newly created VM was started and a message appeared to indicate that the VM had been copied or moved. The newly renamed `.vmdk` file path had to be provided for the VM to start properly. Using the default settings, the VM started successfully and all the data was intact as it originally was on the host machine. The test files installed on the VM were intact and the `md5sum` hashes of the files matched the original test file values.

We encountered two scenarios with respect to deleted VMs. In one instance, the VMware Workstation Ubuntu VM was deleted from the host machine and was partially overwritten by another program (AVG anti-virus log file). In the second instance, another image of the host machine showed the VMware Workstation Ubuntu VM as simply being deleted with no overwriting.

In the first instance of deleting the Ubuntu VM, the entire folder for the VM was still listed in its original location in the host machine file system with all the VM files. The `.vmdk` file was indicated as deleted and was, in fact, overwritten. The `Ubuntu.vmdk` file was copied using the `Copy/UnErase` function as described above. Although the file was overwritten, it was determined (upon searching the data using a hex viewer) that much of the original data still persisted in the file; the overwriting was confined to the beginning of the hard disk file and did not corrupt any operating system information. Since the virtualization hardware is standardized for a given version of VM software (e.g., VMware Workstation 6), recovering partially overwritten data is a possibility. Depending on the position of the lost data in the VM configuration file, it is possible to insert the standardized values to fully recover the configuration file and allow the VM to function. The recovered Ubuntu .vmdk file was opened using Hex Editor Neo. A VM from the analysis machine, Ubuntu Desktop.vmdk, was opened using the same hex editor. The beginning of the Desktop version file, equal to the portion that was determined to be overwritten, was copied and then pasted into the recovered file. This was saved as ubunturepaired.vmdk.

A similar methodology was used to recover and correct the VM configuration file for the deleted Ubuntu VM. The original file was copied from the image via Copy/UnErase and renamed as ubunturepaired.vmx to reflect the naming scheme of the recovered VM. Unnecessary data written to the file (as compared with an intact configuration file) was deleted and the entries were modified to reflect ubunturepaired.vmdk, the new name of the VM virtual hard disk file. This configuration file was renamed as ubunturepaired2.vmx.

Opening the recovered ubunturepaired2.vmx file started the VMware Workstation software. As with the recovered Windows VM, a message was displayed concerning the moving or copying of the VM. Once this message was acknowledged, the recovered Ubuntu VM started and ran normally. All test files were intact and the metadata was visible and correct.

The second instance in which the Ubuntu VM was deleted displayed no traces of the VM in the host machine file system. Searches were conducted to ascertain the location of the .vmdk file. As described above, the file was located and extracted to its own folder on the analysis machine. This was the first step in attempting to recreate the deleted VM. Next, VMware Workstation was run on the analysis machine and a new custom VM was selected for creation. The only difference in creating the VM for recovery was the selection of an existing hard drive instead of creating a new hard drive for the VM. Once this was done, the VM started normally and all the test files were intact. No additional changes were necessary for the VM to run as it was originally configured.

The next step in the recovery process for the completely deleted VM was to extract the configuration file (.vmx). This file was located and extracted to the same folder in an attempt to recreate the conditions under which the original VM was created. The process of recreating the

VM from the second host machine image was not immediately successful using the two extracted files from the unallocated space. Similar to the recreation of the deleted Ubuntu VM, entries were modified in the configuration file to enable it to run the VM. Specifically, the path to the recovered .vmdk file was modified and other extraneous characters copied during the extraction process were removed in order for the configuration file to match the comparison file from the analysis machine.

When the configuration file was opened, VMware Workstation started automatically. The message regarding the moving or copying of the VM was displayed and was acknowledged as being copied. The VM started correctly and all the information was present and intact.

Further research into the process of extracting deleted VMs from the unallocated space on the host machine revealed no discernible differences between the VMs recreated with only the .vmdk file versus those recreated with the .vmdk and .vmx files as well as any other standard file normally created for a new VM. When the extracted/recreated VM was started on the analysis machine, new files were created to complete the file set for the VM, which negated the effects of the other files that were extracted and placed in the VM folder.

8.   Conclusions

The recognition and understanding of the role that virtualization software and VMs play in computing environments in important in digital forensic investigations. Investigators and examiners must be aware that artifacts are produced during the creation and execution of VMs and that many artifacts are recoverable even after the VMs are deleted.

Our research has shown that virtualization software and VMs produce numerous artifacts in a host machine. Specific behavior patterns, such as deleting certain types of virtualization files and even entire VMs, can be used to narrow searches for artifacts. Furthermore, the ability to identify, extract and recreate entire VMs is very useful when investigators and examiners have to determine the circumstances under which the VMs were used.

References

[1] AccessData Corporation, Forensic Toolkit 1.7, Linden, Utah (www.accessdata.com).

[2] F. Bellard, Qemu (bellard.org/qemu).

[3] Canonical, Ubuntu 8.04, London, United Kingdom (www.ubuntu.com).

[4] Citrix Systems, What is Xen? Fort Lauderdale, Florida (www .xen.org).

[5] Digital Intelligence, UltraBlock SATA Bridge Write Blocker, New Berlin, Wisconsin (digitalintelligence.com).

[6] Guidance Software, EnCase 5 and 6, Pasadena, California (guid-ancesoftware.com).

[7] Helios Software Solutions, TextPad, Longridge, United Kingdom (www.textpad.com/index.html).

[8] HHD Software, Free Hex Editor Neo, London, United Kingdom (www.hhdsoftware.com/Products/home/hex-editor-free.html).

[9] Knopper.Net, Knoppix Live Linux Filesystem, Knoppix 5.1.1 Re-lease, Schmalenberg, Germany (www.knopper.net/knoppix/index-en.html).

[10] T. Liston and E. Skoudis, On the cutting edge: Thwarting virtual machine detection (handlers.sans.org/tliston/ThwartingVMDetect ion_Liston_Skoudis.pdf), 2006.

[11] Parallels, Parallels Optimized Computing, Neuhausen am Rheinfall, Switzerland (www.parallels.com).

[12] Parallels, Parallels Workstation 2, Neuhausen am Rheinfall, Switzerland (www.parallels.com/en/products/workstation).

[13] Sun Microsystems, VirtualBox, Santa Clara, California (www.vir tualbox.org).

[14] VMware, VMware, Palo Alto, California (www.vmware.com).

[15] VMware, VMware Workstation 6, Palo Alto, California (www.vm ware.com/products/ws).

[16] VMware, What files make up a virtual machine? Palo Alto, Cal-ifornia (www.vmware.com/support/ws5/doc/ws_learning_files_in_a _vm.html).

[17] Wireshark Foundation, Wireshark, San Jose, California (www.wire shark.org).

[18] X-Ways Software Technology, WinHex, Cologne, Germany (x-ways.net/winhex/index-m.html).

[19] X-Ways Software Technology, X-Ways 14.2, Cologne, Germany (x-ways.net/forensics/index-m.html).

Chapter 6

TEMPORAL ANALYSIS OF WINDOWS MRU REGISTRY KEYS

Yuandong Zhu, Pavel Gladyshev and Joshua James

Abstract   The Microsoft Windows registry is an important resource in digital forensic investigations. It contains information about operating system configuration, installed software and user activity. Several researchers have focused on the forensic analysis of the Windows registry, but a robust method for associating past events with registry data values extracted from Windows restore points is not yet available. This paper proposes a novel algorithm for analyzing the most recently used (MRU) keys found in consecutive snapshots of the Windows registry. The algorithm compares two snapshots of the same MRU key and identifies data values within the key that have been updated in the period between the two snapshots. User activities associated with the newly updated data values can be assumed to have occurred during the period between the two snapshots.

Keywords: MRU registry keys, restore points, registry snapshots

1.   Introduction

The Microsoft Windows registry is "a central hierarchical database" [8] that contains information (stored as keys) related to users, hardware devices and applications installed on the system. As such, it is an important forensic resource that holds a significant amount of information about user activities. This paper focuses on the most recently used (MRU) keys that contain data values (file names, URLs, command line entries, etc.) related to recent user activity [1]. An example is the key `HKCU\Software\Microsoft\Office\12.0\Word\File MRU` that stores the list of recently opened Microsoft Word 2007 documents (Figure 1).

Several MRU keys are used throughout the Windows operating system. Some keys have "MRU" in their names, such as `OpenSaveMRU`,

G. Peterson and S. Shenoi (Eds.): Advances in Digital Forensics V, IFIP AICT 306, pp. 83–93, 2009.

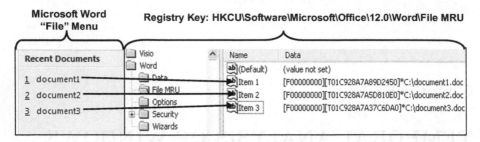

Figure 1. Word file MRU example.

which contains the names of files recently saved by applications that use the standard Microsoft Windows OpenAndSave shell dialog. Others reflect the nature of items in the key such as TypedURLs, which contains a list of the URLs typed into the Internet Explorer address bar by the user. The connections between MRU keys and user actions can be found in several publications (see, e.g., [1, 2, 7]). MRU keys are particularly useful in investigations where it is important to determine the actions performed by specific users [9].

This paper describes an algorithm for analyzing MRU keys in consecutive snapshots of the Windows registry. User activities occurring during the period between the snapshots can be identified by analyzing the updated data values corresponding to a specific MRU key.

2.   Analysis of Registry Snapshots

The Windows registry is stored in the file system in blocks called "hives." System Restore Point hives are backups of the Windows registry created every 24 hours or upon installation of new software [3]. The hives contain several earlier versions of the registry. These versions, which we call "snapshots," can provide an investigator with a detailed picture of how the registry changed over time [4].

Forensic analysis of the registry rarely focuses on the restore points. At best, registry snapshots within the restore points are examined as separate entities during an investigation. This overlooks the links between registry snapshots, which provide more timestamp information about user activities and system behavior than single instances of the Windows registry.

The forensic value of Windows registry information is limited by the relative scarcity of timestamp information [1]. Only one timestamp, the last modification time, is recorded for each registry key [2]. A registry key usually contains multiple data values. Even if it could be determined which value was the last to be updated, it is not possible to determine when the other data values were last modified. This is problematic when an investigator attempts to construct a timeline of events by combining information from several registry keys using only a single instance of the registry.

This problem can be addressed by comparing consecutive registry snapshots and determining which data values were updated. Each registry snapshot is usually associated with a creation time that is recorded as part of the snapshot. Any registry data value updated between two registry snapshots must have been updated between the creation time of the preceding registry snapshot and the last modification time of the registry key that contains the newly updated data value. Furthermore, the activity that caused the registry key update would have occurred within the same time interval.

The YD algorithm presented in this paper compares two snapshots of an MRU key and identifies the data values within the key that were updated during the time interval between the two snapshots. The algorithm provides an investigator with a timeline of changes from the restore points in the Windows registry by reversing the MRU key update process.

3. MRU Key Updates

The obvious way to determine the difference between two MRU key snapshots is to compare them value-by-value and identify the registry data values that were updated. However, although this method is valid for some registry keys, it is not applicable to all MRU keys. As we discuss below, the context in which specific registry data values are stored and the way the values are updated provide clues to the events that took place. In particular, it may be possible to deduce from the context of the data that a specific registry data value was updated even if the content of the data value did not change.

Data updates can be categorized based on two types of MRU keys. The first type of MRU key (Figure 2) stores data in several values that are named using numbers or letters and saves the order of the values in a special value called MRUList or MRUListEx. The leftmost letter of the MRUList or MRUListEx value corresponds to the most recently updated entry in the list. As shown in Figure 2, the value c, which represents the command c:, was the most recently typed command in the Run dialog window in the first snapshot. When a user executes the regedit command, the MRUList value data is updated to the new value sequence and the value a that contains regedit\1 is not affected by this action.

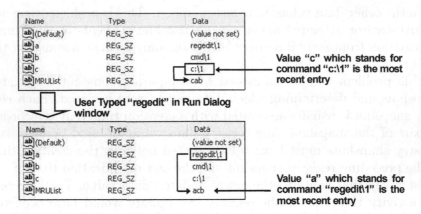

Value "c" which stands for command "c:\1" is the most recent entry

User Typed "regedit" in Run Dialog window

Value "a" which stands for command "regedit\1" is the most recent entry

Figure 2. RunMRU example.

The second type of MRU key consists of a list of most recently used records. However, instead of using a special value to denote the sequence of values in the key, the name of each value contains a number that indicates its order in the list. If the order of values is changed, the system simply renames the values to maintain the order within the key.

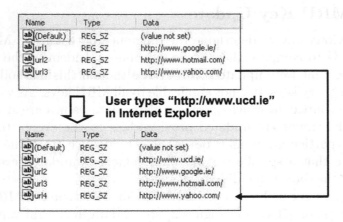

User types "http://www.ucd.ie" in Internet Explorer

Figure 3. TypedURLs example.

Figure 3 presents an example of the second type of MRU key. It shows the consecutive states of the TypedURLs key before and after typing the URL http://www.ucd.ie in Internet Explorer. The value url1 contains the most recently typed URL while url3 contains the least recent entry in the first state. After the new URL http://www.ucd.ie is entered, it is added into the TypedURLs key as a new url1 value. The initial data values in the TypedURLs key are renamed, increasing their index by one – the value url1 becomes url2, url2 becomes url3, and so on.

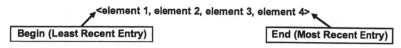

Figure 4. MRU list.

Note that in the example in Figure 2, although the MRU key was updated because `regedit` was typed in the Run dialog window, the corresponding value `a` was not modified. Only the `MRUList` value was updated because the value sequence was changed. On the other hand, all the data values in the second `TypedURLs` key example (Figure 3) were modified even though only one of them (`url1`) was added because of a user action. The `url3` value, for instance, was renamed to `url4` in the new state, but the user action did not change the content. Therefore, when comparing MRU keys in two consecutive snapshots, it cannot be assumed that the content of a changed value within an MRU key was typed, selected or produced by some other user action. The analyst must consider how the particular MRU key was updated in order to decide whether or not a particular value within the key reflects a user action that occurred between two states of the MRU key.

3.1   MRU List

Although different update processes are used for the two types of MRU keys, the keys are still updated in a similar manner. Consequently, a common model – the ordered MRU list – is used to represent the different MRU key types in the registry. As shown in Figure 4, the first element of the MRU list corresponds to the oldest entry of the MRU key; the last element of the list is the most recent entry of the MRU key. The MRU list thus abstracts the details of the MRU update process.

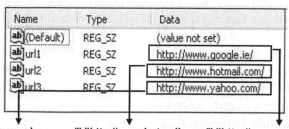

<"http://www.yahoo.com/","http://www.hotmail.com/","http://www.google.ie/">

Figure 5. `TypedURLs` MRU list.

Figures 5 and 6 illustrate how the two types of MRU keys are converted into MRU lists.

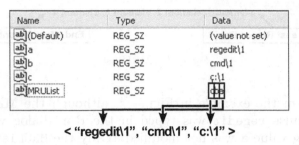

Figure 6. RunMRU MRU List.

Since most MRU keys have a limit on the number of elements, their MRU lists inherit this property. The maximum number of elements varies according to the key. For example, TypedURLs can have at most 25 elements while OpenSaveMRU can store at most ten elements. The least recent entry is removed when the number of elements in the list exceeds the maximum allowed.

3.2   MRU List Updating Algorithm

Updating the MRU list involves two steps [5, 6]. The first is common to every MRU list; it describes how a new element is added to the existing MRU list. The second involves additional list processing that is unique to each MRU key.

The updating process begins when the system receives a request to add a new element to the MRU list. The first step enumerates the existing MRU list and compares each entry with the new element. If no entries match, the new element is appended to the end of the current MRU list. If a match occurs, the old entry is removed and the new element is appended to the end of MRU list.

In the second step, the new list is processed based on additional constraints imposed by the particular MRU list. For most MRU lists, the constraint is the limit on the maximum number of elements (which results in the least recent element being removed when the maximum is exceeded). However, some MRU lists have an additional constraint, which causes elements to be removed from the list even when the maximum number of elements is not exceeded.

Figure 7 presents three examples of the MRU list updating process. The MRU list is assumed to have a maximum size of five. The first example adds a new element 5.txt to the existing MRU list <1.txt, 2.txt, 3.txt, 4.txt, 5.txt>. The system detects that 5.txt is already in the list; therefore, 5.txt is removed upon which the new 5.txt element is added to the list. The same process is followed when a new

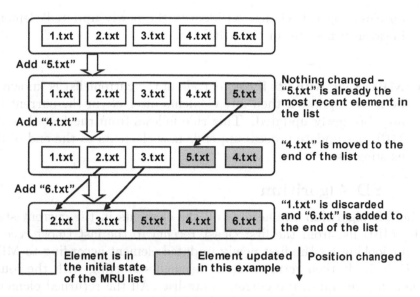

Figure 7. MRU list updating example.

4.txt item is added; the old 4.txt is removed and the new 4.txt is placed at the end of the list. In the third example, a new element 6.txt is added to the current MRU list. Because 6.txt is not already in the list, the system appends 6.txt to the end of the list. However, the new list exceeds the maximum number of elements, causing the oldest item in the list (1.txt) to be removed.

3.3   MRU List Update Rules

Based on the list updating process, we specify two rules for identifying the newly updated elements between two states of the MRU list.

- **MRU List Rule 1:** Element *ele* is a newly updated element if there exists an element before *ele* in the current state of the MRU list that does not appear before *ele* in the previous state of the MRU list. To understand this rule, consider the second step in Figure 7 where 4.txt is added to the list. In the state that immediately precedes this step, the element 4.txt is preceded by the elements 1.txt, 2.txt and 3.txt in the list. However, when 4.txt is "added" to the MRU list, the element 4.txt is moved to the end of the list, causing it to be preceded by 5.txt in addition to 1.txt, 2.txt and 3.txt. Note that the "happened before" relationship between 4.txt and 5.txt has changed. The rule relies on detecting this change to identify the newly updated elements between two states of the MRU list.

- **MRU List Rule 2**: If element *ele* in the current list is known to be newly updated, then any elements after *ele* in the current list are also newly updated. This rule follows from the fact that the MRU list update process appends new elements to the end of the existing list.

4.   YD Algorithm

The YD algorithm enumerates all the elements in the current state of the MRU list from the first (least recent) to the last (most recent) element looking for the first newly updated element according to MRU List Rule 1. It then returns all the elements starting with the found element to the end of the current state list. All the returned elements are identified as newly updated according to the MRU List Rule 2. Since an MRU list is a model of a particular MRU key, the newly updated elements identidied in the MRU list correspond to the newly updated data values in the MRU key.

Figure 8 presents the flowchart of the YD algorithm. List A denotes the previous state of the MRU list with n elements and List B is the current state of the MRU list with m elements. Note that A_x and B_y denote the xth and y^{th} elements of A and B, respectively.

The YD algorithm first checks if A or B are empty. If B is empty, the algorithm returns NULL. On the other hand, if A is empty, the algorithm returns the elements of B. Next, the algorithm compares the first element B_1 of (current) List B with each element of (previous) List A. The loop ends either when an element A_x is equal to B_1 or the end of List A is encountered. If a match for B_1 is not found, the algorithm terminates and returns the entire List B. If B_1 is found in A, the algorithm continues to compare each consecutive element in B with each element in A following the matched A_x element. The process terminates when element B_y is not found in A (and the algorithm returns the elements from B_y to B_m) or the algorithm reaches the end of List B (and the algorithm returns NULL).

Figure 9 shows an example involving the first and last MRU lists from Figure 7. The algorithm requires three steps to identify the newly updated elements 4.txt and 6.txt. These results match the results of the updating process discussed in Section 3.2. Note, however, that one updated element 5.txt is not identified. MRU List Rule 1 does not detect 5.txt as newly updated because the addition of 5.txt to the

Assume: List A is the previous state of the MRU list with *n* elements and B is the current state of the MRU list that contains *m* elements

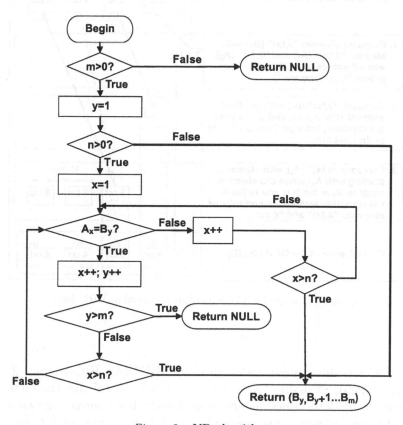

Figure 8. YD algorithm.

MRU list does not change the "happened before" order of the MRU list elements.

5. YD Algorithm Limitations

As discussed above, the YD algorithm may not detect all the elements updated between two MRU lists. This is because the algorithm relies on two (somewhat limited) rules that rely on the fact that elements change their positions in the list when they are updated. However, in some cases (as in the first step of the updating example in Figure 7), the state of the MRU key does not change in response to a user action.

To address this issue, we propose that the elements in the most recent MRU list be divided into a "Definitely Newly Updated Elements" set and a "Possibly Newly Updated Elements" set. "Definitely Newly Updated

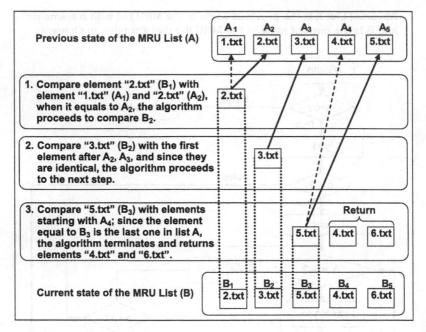

Figure 9.   Using the YD algorithm to examine MRU lists.

Elements" are the elements identified by MRU List Rules 1 and 2. On the other hand, "Possibly Newly Updated Elements" are the elements that cannot be determined as newly updated when comparing two MRU lists. In the example discussed above, 2.txt, 3.txt and 5.txt are possibly newly updated elements – it cannot be determined from MRU data alone if these elements were actually updated. However, if some external evidence exists to suggest that one of the three elements (say 3.txt) was newly updated, then, according to MRU List Rule 2, any possibly newly updated elements that happened after it (i.e., 5.txt) must also have been updated during the same time period.

6.   Conclusions

The YD algorithm for analyzing MRU keys found in different snapshots of the Windows registry relies on two rules derived from a generic MRU key update model. Although the algorithm may not detect all the newly updated data values corresponding to MRU keys, it can identify "definitely newly updated" values. These data values are important because they enable investigators to determine the associated events that occurred between the two snapshots. The algorithm can be used to examine MRU keys in consecutive registry snapshots extracted from Windows restore points as well as from consecutive backups of a system.

Acknowledgements

This research was funded by the Science Foundation of Ireland under the Research Frontiers Program 2007 Grant No. CMSF575.

References

[1] H. Carvey, The Windows registry as a forensic resource, *Digital Investigation*, vol. 2(3), pp. 201–205, 2005.

[2] H. Carvey, *Windows Forensic Analysis*, Syngress, Burlington, Massachusetts, 2007.

[3] B. Harder, Microsoft Windows XP system restore, Microsoft Corporation, Redmond, Washington (technet.microsoft.com/en-us/lib rary/ms997627.aspx), 2001.

[4] K. Harms, Forensic analysis of system restore points in Microsoft Windows XP, *Digital Investigation*, vol. 3(3), pp. 151–158, 2006.

[5] J. Holderness, MRU lists (Windows 95) (www.geocities.com/Silicon Valley/4942/mrulist.html), 1998.

[6] E. Kohl and J. Schmied, `comctl32undoc.c`, Wine Cross Reference (source.winehq.org/source/dlls/comctl32/comctl32undoc.c), 2000.

[7] V. Mee, T. Tryfonas and I. Sutherland, The Windows registry as a forensic artifact: Illustrating evidence collection for Internet usage, *Digital Investigation*, vol. 3(3), pp. 166–173, 2006.

[8] Microsoft Corporation, Windows registry information for advanced users, Redmond, Washington (support.microsoft.com/kb/256986), 2008.

[9] B. Sheldon, Forensic analysis of Windows systems, in *Handbook of Computer Crime Investigation: Forensic Tools and Technology*, E. Casey (Ed.), Academic Press, London, United Kingdom, pp. 133–166, 2002.

Chapter 7

USING DCT FEATURES FOR PRINTING TECHNIQUE AND COPY DETECTION

Christian Schulze, Marco Schreyer, Armin Stahl and Thomas Breuel

Abstract   The ability to discriminate between original documents and their photocopies poses a problem when conducting automated forensic examinations of large numbers of confiscated documents. This paper describes a novel frequency domain approach for printing technique and copy detection of scanned document images. Tests using a dataset consisting of 49 laser-printed, 14 inkjet-printed and 46 photocopied documents demonstrate that the approach outperforms existing spatial domain methods for image resolutions exceeding 200 dpi. An increase in classification accuracy of approximately 5% is achieved for low scan resolutions of 300 dpi and 400 dpi. In addition, the approach has the advantage of increased processing speed.

Keywords: Printing technique and copy detection, discrete cosine transformation

1.   Introduction

Due to advances in digital imaging techniques, it is relatively simple to create forgeries or alter documents within a short timeframe. According to the American Society of Questioned Document Examiners (ASQDE), modern printing technologies are increasingly used to produce counterfeit banknotes [5] and forged documents [13].

Forensic document examiners are confronted with a variety of questions [9]: Who created the document? Which device created the document? What changes have been made to the document since it was originally produced? Is the document as old as it purports to be? As a result, a variety of sophisticated methods and techniques have been developed since Osborn and Osborn [16] noted that a document may have any one of twenty or more different defects that may not be observed unless one is specifically looking for them.

G. Peterson and S. Shenoi (Eds.): Advances in Digital Forensics V, IFIP AICT 306, pp. 95–106, 2009.

One challenge when examining machine-printed documents is to distinguish between laser-printed and photocopied documents. Kelly, *et al.* [9] note that "the analysis of a photocopy can run the entire gamut of instruments in a document examination laboratory." The difficulty can be ascribed to the fact that laser printing and photocopying are very similar in operation [6] – both techniques use indirect electrostatic digital imaging to transfer the printing substrate.

The photocopy process has at least two distinct phases: scanning the template document and printing the scanned content. Due to technical limitations of photocopier devices and/or physical conditions, a small amount of the original document information may be lost or altered. These imperfections are observable at the edges of characters in the form of light blurring [3]. The imperfections are more evident after the document is transformed into the frequency domain. This is because, in general, sharp transitions between character and non-character areas result in a larger number of high frequency values compared with the smooth transitions associated with blurred character edges.

This work presents a novel frequency domain approach for document printing technique recognition. The recognition system creates a unique fingerprint for each printing technology from the number and distribution of the frequencies contained in a document. Experimental results demonstrate that using discrete cosine transformation (DCT) coefficients and machine learning techniques can help distinguish between inkjet-printed and laser-printed documents and also detect first-generation photocopies at low scan resolutions.

2.   Related Work

Kelly and Lindblom [9] have surveyed the major forensic document techniques applicable to questioned documents. Although the use of digital imaging techniques in forensic examinations of documents is relatively new, recent research efforts have demonstrated that these techniques are very useful for discriminating between non-impact printing techniques.

Print quality is a useful feature for identifying the printing technology used to create a document. Print quality metrics include line width, raggedness and over spray, dot roundness, perimeter and the number of satellite drops [15]. Other useful quality features include the gray level co-occurrence features for the printed letter "e" [14]; and texture features based on gray level co-occurrence and local binary map histograms used in conjunction with edge roughness features that measure line edge roughness, correlation and area difference for printed document characters [11]. Schulze, *et al.* [17] have evaluated these features in the context of high-throughput document management systems.

The methodology proposed by Tchan [18] is very similar to our approach. Documents are captured at low resolution and printing technologies are distinguished by measuring edge sharpness, surface roughness and image contrast. However, Tchan has only experimented with documents containing squares and circles, not typical office documents.

Color laser printers add another dimension to the document feature space, which has led to the development of alternative detection methods. By evaluating hue component values within the HSV color space, it is possible to distinguish between different printing substrates and, thus, printing techniques [6]. Additionally, the yellow dot protection patterns on documents printed by color laser printers (that are nearly invisible to the unaided human eye) can be used for printer identification [12, 19]. Indeed, the distinctive dot pattern is directly related to the serial number of a particular laser printer or photocopier.

In addition to the printing technology, the physical characteristics of a printing device often leave distinctive fingerprints on printed documents. For example, the manner in which the spur gears hold and pass paper through printing devices can be used to link questioned documents to suspected printers [2].

Gupta, *et al.* [7] have presented a structured methodology for detecting the scanner-printer combination used to create tampered documents based on document image imperfections. They measure the overall similarity and the similarity in coarse document areas between the original document and the tampered document. The standard deviation and average saturation of the image noise are then computed. Experimental results indicate that this method is very effective.

However, none of the techniques in the literature use frequency domain features to identify the printing technique used to create a questioned document. Also, we were unable to find any research articles focusing on the detection of photocopied documents in real-world scenarios.

3.   Printing Process Characteristics

In general, printing is a complex reproduction process in which ink is applied to a printing substrate in order to transmit information in a repeatable form using image-carrying media (e.g., a printing plate) [10].

Inkjet printers use a printhead to emit tiny ink droplets onto the printing paper. As the paper is drawn through the printer, the printhead moves back-and-forth horizontally and usually transfers the ink directly to the paper. Ink deposition is digitally controlled by the printer

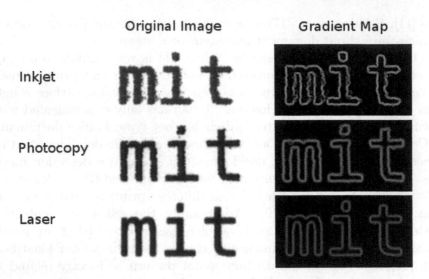

Figure 1.   Representative plots of printing technologies.

firmware. Ink is sprayed onto the paper so that multiple gradients of dots accumulate to form an image with variable color tones.

The majority of laser printers and photocopiers engage electrophotographic printing technology. The underlying concept is to generate a visible image using an electrostatic latent image created by surface charge patterns on a photoconductive surface. The main difference between laser printing and photocopying is the image source. In the case of photocopying, the image has to be scanned prior to printing. During the scanning procedure, a scanhead is moved slowly across the illuminated document. The scanhead divides the image into microscopic rows and columns and measures how much light reflects from each individual row-column intersection. The charge collected by each photodiode during the scanning process is proportional to the reflection of a specific area of the paper. The amount of reflectance is then recorded as a dot or picture element (pixel).

Edge sharpness is the most effective feature for visually discriminating between high resolution scans of documents created by different printing techniques. Figure 1 shows representative plots of inkjet, photocopy and laser printing technologies and their corresponding gradient maps. The images are captured at a scanning resolution of 2,400 dpi. The left column shows the original scans while the right column displays the corresponding gradient maps obtained using a Prewitt gradient filter. Note the differences in edge sharpness and edge roughness for the different printing techniques.

3.1   Edge Sharpness and Contrast

The gradient at an image pixel measures the image intensity change in the direction of the highest intensity change at that particular pixel. A sharp edge and, therefore, a high intensity change results in a skin gradient. Examination of the gradient images for the three printing techniques in Figure 1 reveals that the laser-printed document image is characterized by sharp transitions between character and non-character areas. In contrast, the photocopied and the inkjet-printed images show a tendency towards smoother and blurred character edges. This feature is due to printing substrate diffusion in the case of the inkjet-printed image and light diffusion during scanning in the case of the photocopied document.

3.2   Edge Roughness and Degradation

Edge roughness and degradation denote the divergence of the printed character shape from the original template character shape. The gradient maps in Figure 1 reveal different degrees of character shape degradation. A high degree of edge roughness is observed for the inkjet-printed document; in contrast, little edge roughness is seen for the laser-printed document. As with edge sharpness, edge roughness is determined by several factors such as printer resolution, dot placement accuracy, rendering algorithms and interactions between colorant and paper.

4.   Frequency Domain Comparisons

Figure 2 shows the results of a pairwise printing technique comparison in the frequency domain. The DCT [1] is first calculated for images produced by each document creation technique; the DCT frequency spectrum results are then averaged for each document creation technique and compared. The left image shows the result obtained by comparing the average laser-printed spectrum (white) versus the average spectrum of the photocopied documents (black). The middle image shows the average laser-printed spectrum (white) versus the average inkjet-printed spectrum (black). The right image shows the average inkjet-printed spectrum (white) versus the average spectrum of the photocopied documents (black). The frequency comparison spectra images in Figure 2 reveal clear differences between the techniques. Comparing the average DCT coefficient spectrum of laser-printed documents with those produced by the other two printing technologies shows a radial symmetric pattern. This pattern is not as distinctive, but is still recognizable when comparing the spectra of the inkjet-printed and photocopied documents.

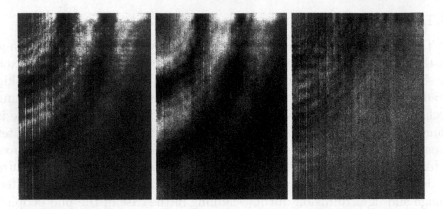

Figure 2.   Frequency domain comparison of printing technique classes.

The difference can be traced to the relationship between the spatial and frequency domains. Sharp edges in the spatial domain yield increased DCT coefficient values for high frequencies. Compared with characters in photocopied and inkjet-printed documents, laser-printed characters have sharper transitions between character and non-character regions. This property is evident in the DCT coefficient values.

Comparison of the images in Figure 2 also indicates that the DCT coefficients corresponding to horizontal and vertical frequencies exhibit a highly discriminant behavior. This is due to the properties of Latin fonts for which large numbers of sharp transitions exist between character and non-character regions in the horizontal and vertical directions. These sharp transitions are very fragile to edge blurring induced by inkjet printing and photocopying. As a result, both printing techniques show a tendency to smaller coefficient values in the high frequencies of the horizontal and vertical components of the DCT spectrum.

5.   DCT Coefficient Distribution Analysis

The frequency difference images in Figure 2 and the printing technique characteristics discussed above underscore the idea that printing techniques are distinguishable according to their frequency spectra. This especially holds for frequency sub-band coefficients corresponding to vertical and horizontal image intensity variations in the spatial domain.

A two-step procedure is used to determine the DCT coefficient distribution and the distribution strength within the frequency spectrum sub-bands for an arbitrary document image. First, the DCT coefficients of a particular sub-band are extracted from the frequency spectrum. Next, statistical features are calculated from the sub-band coefficients.

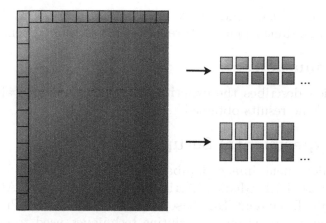

Figure 3. Frequency sub-band coefficient detection and extraction.

- **Frequency Sub-Band Extraction:** As shown in Figure 2, a high discriminative character is evident in the horizontal and vertical frequency sub-bands of the spectra. Therefore, these frequency sub-band coefficients are obtained from the spectrum of each document by extracting a set of horizontal and vertical sub-band boxes as illustrated in Figure 3. The left-hand side figure shows the detected horizontal and vertical frequency sub-band coefficients; the right-hand side shows the extracted frequency sub-band coefficients. The aspect ratio of the sub-band boxes is set to $\frac{1}{\sqrt{2}}$ in order to capture the frequency distortion caused by the aspect ratio of the document images.

- **Statistical Feature Extraction:** Let k be the number of frequency sub-band boxes box_i obtained from the normalized frequency spectrum $F'(u, v)$ of a document. To obtain the unique horizontal and vertical frequency sub-band pattern of the document, the mean and standard deviation of the coefficients are computed according to the following equations for each sub-band box_i:

$$\mu(box_i) = \frac{1}{MN} \sum_{m=0}^{M} \sum_{n=0}^{N} F'(m, n)$$

$$\sigma(box_i) = \left(\frac{1}{MN} \sum_{m=0}^{M} \sum_{n=0}^{N} (F'(m, n) - \mu)^2 \right)^{\frac{1}{2}}$$

where $m, n \in box_i$ and M, N indicate the size of box_i. This produces a $2k$-dimensional feature vector for each document image.

6.   Evaluation

This section describes the experimental setup used to evaluate our approach and the results obtained.

6.1   Experimental Setup

Existing document image databases (e.g., UW English Document Image Database I-III, Medical Article Records System (MARS), MediaTeam Oulu Document Database, Google 1,000 Books Project) do not include annotations on the printing techniques used to create the documents. Therefore, a document image database annotated with the needed ground truth information was created. As in previous works [11, 17], the "Grünert" letter in the 12 pt "Courier New" normal font with a 12 pt line height was used as a template to create the ground truth document image database. This template document implements the DIN-ISO 1051 standard. The database used consisted of 49 laser and 14 inkjet printouts and 46 photocopied documents. Every document in the database was created by a different printer or photocopier, covering all the major manufacturers.

To create a realistic evaluation scenario, half of the photocopied documents were generated using laser-printed templates while the other half were based on inkjet-printed templates. Only first-generation photocopies were added to the database.

All the documents were scanned at resolutions of 100 dpi, 200 dpi, 300 dpi, 400 dpi and 800 dpi. A Fujitsu 4860 high-speed scanning device was used for scan resolutions lower than 400 dpi. This scanner is designed for high-throughput scanning and, therefore, the maximal scan resolution is limited to 400 dpi. In order to test the DCT feature performance at higher resolutions, an EPSON 4180 device was used to produce 800 dpi scans. All the document images obtained were stored in the TIFF file format to avoid further information loss.

Classification Methodology   We classified the printing techniques of the scanned document images using support vector classification utilities provided by the LibSVM library [4]. The support vector machine (SVM) classification was performed using a radial-basis function kernel with optimized parameters. Optimal kernel parameters for C and γ were obtained by a coarse grid search in the SVM parameter space within the intervals $C = [2^{-5}, 2^{15}]$ and $\gamma = [2^{-15}, 2^3]$ as suggested in [8].

Table 1. Classification accuracy results.

dpi	DCT [%]	v_{DCT} $[\frac{p}{min}]$	Gradient [%]	v_{grad} $[\frac{p}{min}]$
100	72.64	44.78	75.47	50.85
200	80.95	15.50	80.00	14.60
300	85.85	6.42	81.13	6.26
400	92.92	2.22	88.23	2.01
800	99.08	0.53	97.17	0.78

Performance Evaluation To evaluate the prediction capability of the extracted features without losing the generalization ability of the learned model, we applied a stratified 10-fold cross validation. To evaluate the classification performance of the applied feature data, we calculated (for each classification trial) the accuracy based on the percentage of correctly classified documents in the testing data set. The average accuracy across all trials was then computed to give the overall result of the stratified cross-validation. The results were compared with a gradient feature based on the work of Tchan [18]. We selected this feature because it was the best performing feature from the set of implemented spatial domain features.

6.2   Experimental Results

Table 1 presents the classification accuracy results for printing techniques using the DCT feature and the best performing spatial domain feature. The processing speed is given in pages per minute. The values in Table 1 show that the DCT feature produces better classification accuracy rates than the gradient feature (the best performing spatial domain feature) for scan resolutions $\geq$ 200 dpi. Significantly better performance of $\approx 5\%$ is seen with the DCT feature for resolutions of 300 dpi and 400 dpi. This observation is important for high-throughput systems because the maximal scan resolution of these systems is usually limited to 400 dpi. Note that the classification accuracy of the DCT feature for the three classes (inkjet, laser and copy) at 400 dpi exceeds 90%.

Figure 4 presents the classification accuracy (top) and processing speed (bottom) for DCT and gradient-based features. The throughput achieved for resolutions of 200 to 400 dpi is larger for the DCT feature, while at 100 dpi and 800 dpi the gradient-based feature has larger values. The speed measurements were made on a system equipped with an Intel Core 2 Duo T7300 (2 GHz) processor and 1 GB memory using a single core.

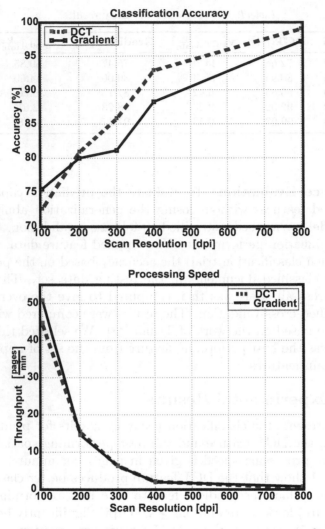

Figure 4. Classification accuracy and processing speed.

7. Conclusions

This paper has presented a novel approach for document printing technique recognition using features calculated in the frequency domain. It
demonstrates that the frequency distribution and number of frequencies
within a transformed document image are directly related to the edge
and noise characteristics visible in the spatial domain. Therefore, the
DCT coefficients obtained by transforming a document image can serve
as a fingerprint for the printing technology used to create the document.

Experimental results verify that the frequency domain approach outperforms spatial domain techniques. The results are particularly significant in the case of document images scanned at resolutions from 200 dpi to 800 dpi. Moreover, no major increase in the processing time is observed despite the fact that a transformation of the scanned document into the frequency domain is necessary to extract DCT based features. Consequently, the approach presented in this paper is also well-suited to high-throughput document processing scenarios.

References

[1] N. Ahmed, T. Natarajan and K. Rao, Discrete cosine transform, *IEEE Transactions on Computers*, vol. 23(1), pp. 90–93, 1974.

[2] Y. Akao, K. Kobayashi and Y. Seki, Examination of spur marks found on inkjet-printed documents, *Journal of Forensic Science*, vol. 50(4), pp. 915–923, 2005.

[3] H. Baird, The state of the art of document image degradation modeling, *Proceedings of the Fourth International Association for Pattern Recognition Workshop on Document Analysis Systems*, pp. 1–16, 2000.

[4] C. Chang and C. Lin, LIBSVM: A Library for Support Vector Machines, Department of Computer Science and Information Engineering, National Taiwan University, Taipei, Taiwan (www.csie .ntu.edu.tw/~cjlin/libsvm).

[5] J. Chim, C. Li, N. Poon and S. Leung, Examination of counterfeit banknotes printed by all-in-one color inkjet printers, *Journal of the American Society of Questioned Document Examiners*, vol. 7(2), pp. 69–75, 2004.

[6] H. Dasari and C. Bhagvati, Identification of printing process using HSV colour space, *Proceedings of the Seventh Asian Conference on Computer Vision*, pp. 692–701, 2006.

[7] G. Gupta, R. Sultania, S. Mondal, S. Saha and B. Chanda, A structured approach to detect the scanner-printer used in generating fake documents, *Proceedings of the Third International Conference on Information Systems Security*, pp. 250–253, 2007.

[8] C. Hsu, C. Chang and C. Lin, A Practical Guide to Support Vector Classification, Department of Computer Science and Information Engineering, National Taiwan University, Taipei, Taiwan (www .csie.ntu.edu.tw/~cjlin/papers/guide/guide.pdf), 2003.

[9] J. Kelly and B. Lindblom (Eds.), *Scientific Examination of Questioned Documents*, CRC Press, Boca Raton, Florida, 2006.

[10] H. Kipphan, *Handbook of Print Media*, Springer, Heidelberg, Germany, 2001.

[11] C. Lampert, L. Mei and T. Breuel, Printing technique classification for document counterfeit detection, *Proceedings of the International Conference on Computational Intelligence and Security*, pp. 639–644, 2006.

[12] C. Li, W. Chan, Y. Cheng and S. Leung, The differentiation of color laser printers, *Journal of the American Society of Questioned Document Examiners*, vol. 7(2), pp. 105–109, 2004.

[13] J. Makris, S. Krezias and V. Athanasopoulou, Examination of newspapers, *Journal of the American Society of Questioned Document Examiners*, vol. 9(2), pp. 71–75, 2006.

[14] A. Mikkilineni, P. Chiang, G. Ali, G. Chiu, J. Allebach and E. Delp, Printer identification based on gray level co-occurrence features for security and forensic applications, *Proceedings of the SPIE*, vol. 5681, pages 430–440, 2005.

[15] J. Oliver and J. Chen, Use of signature analysis to discriminate digital printing technologies, *Proceedings of the International Conference on Digital Printing Technologies*, pp. 218–222, 2002.

[16] A. Osborn and A. Osborn, Questioned documents, *Journal of the American Society of Questioned Document Examiners*, vol. 5(1), pp. 39–44, 2002.

[17] C. Schulze, M. Schreyer, A. Stahl and T. Breuel, Evaluation of gray level features for printing technique classification in high-throughput document management systems, *Proceedings of the Second International Workshop on Computational Forensics*, pp. 35–46, 2008.

[18] J. Tchan, The development of an image analysis system that can detect fraudulent alterations made to printed images, *Proceedings of the SPIE*, vol. 5310, pp. 151–159, 2004.

[19] J. Tweedy, Class characteristics of counterfeit protection system codes of color laser copiers, *Journal of the American Society of Questioned Document Examiners*, vol. 4(2), pp. 53–66, 2001.

Chapter 8

SOURCE CAMERA IDENTIFICATION USING SUPPORT VECTOR MACHINES

Bo Wang, Xiangwei Kong and Xingang You

Abstract   Source camera identification is an important branch of image forensics. This paper describes a novel method for determining image origin based on color filter array (CFA) interpolation coefficient estimation. To reduce the perturbations introduced by a double JPEG compression, a covariance matrix is used to estimate the CFA interpolation coefficients. The classifier incorporates a combination of one-class and multi-class support vector machines to identify camera models as well as outliers that are not in the training set. Classification experiments demonstrate that the method is both accurate and robust for double-compressed JPEG images.

Keywords: Camera identification, CFA interpolation, support vector machine

1. Introduction

Sophisticated digital cameras and image editing software increase the difficulty of verifying the integrity and authenticity of digital images. This can undermine the credibility of digital images presented as evidence in court. Two solutions exist, watermarking and digital image forensics. Compared with the active approach of digital watermarking, digital image forensics [7, 12] is a more practical, albeit more challenging, approach. In a digital forensics scenario, an analyst is provided with digital images and has to gather clues and evidence from the images without access to the device that created them [15]. An important piece of evidence is the identity of the source camera.

Previous research on source camera identification has focused on detecting defective sensor points [5] and generating reference noise patterns for digital cameras [10]. The reference noise pattern for a digital camera is obtained by averaging over a number of unprocessed images. The

G. Peterson and S. Shenoi (Eds.): Advances in Digital Forensics V, IFIP AICT 306, pp. 107–118, 2009.

source camera corresponding to an image is identified using a correlator between the reference pattern noise and the noise extracted from the image. These methods suffer from the limitation that the analyst needs access to the digital camera to construct the reference pattern. Moreover, the reference pattern is camera-specific instead of model-specific.

Several methods have been proposed for identifying the source camera model. These methods primarily extract features from the digital image and use a classifier to determine image origin. The method of Kharrazi, *et al.* [8] uses image color characteristics, image quality metrics and the mean of wavelet coefficients as features for classification. Although this method has an average classification accuracy of nearly 92% for six different cameras, it cannot easily distinguish between cameras of the same brand but different models. The classification accuracy can be improved by combining the feature vector in [8] with the lens radial distortion coefficients of digital cameras [4]. However, extracting distorted line segments to estimate the distortion parameters limits the application of this method to images that contain distorted line segments. Meanwhile, good performance has been obtained by combining bi-coherence and wavelet features in a classifier [11].

Recently, several algorithms that use color filter array (CFA) interpolation coefficients have been developed. Most digital cameras use a number of sensors to capture a mosaic image, where each sensor senses only one color – red (R), green (G) or blue ((B). Consequently, a CFA interpolation operation called "demosaicking" is necessary to obtain an RGB color image. A variety of CFA interpolation patterns are used; the most common is the Bayer pattern. Bayram, *et al.* [1] employ an expectation-maximization algorithm to extract the spectral relationship introduced by interpolation to build a camera-brand classifier. Long and Huang [9] and Swaminathan, *et al.* [14, 15] have developed CFA interpolation coefficient estimation methods based on the quadratic pixel correlation model and the minimization problem. The best experimental results were obtained by Swaminathan, *et al.* [15], who achieved an average classification accuracy of 86% for nineteen camera models.

Most of the methods discussed above use Fisher's linear discriminant or support vector machine (SVM) classifiers. But these classifiers only distinguish between classes included in their training model – a false classification occurs when an item belonging to a new class is presented. Another problem is that the images for source camera identification often have double JPEG compression, which usually causes the methods discussed above to incorrectly classify the images.

This paper focuses on the important problem of identifying the camera source of double-compressed JPEG images. The method addresses the difficulties posed by outlier camera model detection and identification. A classifier that combines one-class and multi-class SVMs is used to distinguish between outlier camera models. The image features use the covariance matrix to estimate the CFA interpolation coefficients used to accurately identify the source camera model. Experimental results based on sixteen different camera models demonstrate the robustness of the approach.

2.   CFA Coefficient Features

A CFA interpolation algorithm, which is an important component of the imaging pipeline, leaves a unique pattern on a digital image. Such an algorithm is brand- and often model-specific. Consequently, CFA coefficients derived from an image can be used to determine image origin. An accurate estimation of the CFA coefficients improves classification accuracy. Our method applies the covariance matrix to reduce the negative impact of JPEG compression in the linear CFA interpolation model when the coefficients are estimated.

Practically every CFA interpolation algorithm interpolates missing RGB pixels in a mosaic image from a small local neighborhood. Thus, the interpolation operation can be modeled as a weighted linear combination of neighbor pixels in RGB channels [1, 14, 15]. For example, a missing G pixel $g_{x,y}$ is interpolated using an $n \times n$ neighborhood as:

$$g_{x,y} = \left. \sum_{i=-n}^{n} \sum_{j=-n}^{n} w_g g_{x+i,y+j} \right|_{\substack{except \\ i=0 \& j=0}} + \sum_{i=-n}^{n} \sum_{j=-n}^{n} w_r r_{x+i,y+j}$$
$$+ \sum_{i=-n}^{n} \sum_{j=-n}^{n} w_b b_{x+i,y+j}$$

where w_g, w_r and w_b are the weighted coefficients in the interpolation.

The linear model can be expressed in vector form as:

$$p = [\vec{W_r} \quad \vec{W_g} \quad \vec{W_b}] * \begin{bmatrix} \vec{R} \\ \vec{G} \\ \vec{B} \end{bmatrix}$$

where p is the interpolated value and $\vec{R}$, $\vec{G}$ and $\vec{B}$ refer to the R, G and B pixel values, respectively, whose center is the interpolated pixel.

For an image with $M \times N$ resolution, each interpolation operation can be described as:

$$p_k = \sum_{l=1}^{3n^2-1} w_l s_{l,k}, \quad k \in [1, M \times N]$$

where s_{lk} denotes the $n \times n$ pixel values of the three channels except the kth interpolated value, and w_l is the corresponding interpolation coefficient weight.

Equivalently, the vector expression $\vec{P} = \vec{W} * \vec{S}$ can be written as:

$$\vec{P} = \begin{bmatrix} p_1 \\ p_2 \\ \vdots \\ p_{M \times N} \end{bmatrix}$$

$$= \begin{bmatrix} w_1 s_{1,1} + w_2 s_{2,1} + \ldots + w_{3n^2-1} s_{3n^2-1,1} \\ w_2 s_{1,2} + w_2 s_{2,2} + \ldots + w_{3n^2-1} s_{3n^2-1,2} \\ \vdots \\ w_1 s_{1,M \times N} + w_2 s_{2,M \times N} + \ldots + w_{3n^2-1} s_{3n^2-1,M \times N} \end{bmatrix}$$

$$= \vec{W} * \vec{S}$$

JPEG compression is a common post-processing operation used in image storage that follows CFA interpolation in the imaging pipeline. An additional JPEG compression to reduce file size is commonly performed when an image is intended to be distributed over the Internet. A JPEG compression is lossy and alters the pixel values from the CFA interpolated results. To counter this, we introduce a term in each interpolation to model the perturbation introduced by single and double JPEG compressions:

$$\vec{P}' = \vec{P} + \vec{\delta} = \begin{bmatrix} p_1 \\ p_2 \\ \vdots \\ p_{M \times N} \end{bmatrix} + \begin{bmatrix} \delta_1 \\ \delta_2 \\ \vdots \\ \delta_{M \times N} \end{bmatrix}$$

$$= \begin{bmatrix} w_1 s_{1,1} + w_2 s_{2,1} + \ldots + w_{3n^2-1} s_{3n^2-1,1} + \delta_1 \\ w_2 s_{1,2} + w_2 s_{2,2} + \ldots + w_{3n^2-1} s_{3n^2-1,2} + \delta_2 \\ \vdots \\ w_1 s_{1,M \times N} + w_2 s_{2,M \times N} + \ldots + w_{3n^2-1} s_{3n^2-1,M \times N} + \delta_{M \times N} \end{bmatrix}$$

This can be written as:

$$\vec{P}' = w_1 \vec{S}_{1,k} + w_2 \vec{S}_{2,k} + \ldots + w_{3n^2-1} \vec{S}_{3n^2-1,k} + \vec{\delta}$$

where $\vec{S}_{l,k} = [s_{l,1} \quad s_{l,2} \quad \cdots \quad s_{l,M \times N}]'$, $l \in [1, 3n^2 - 1]$ is the vector of pixel values in the neighborhood of the interpolated location. In this formulation, we attempt to estimate all the interpolation coefficients w_l using the covariance between $\vec{P}'$ and $\vec{S}_{l,k}$:

$$
\begin{aligned}
cov(\vec{P}', \vec{S}_{l,k}) &= cov(w_1 \vec{S}_{1,k} + w_2 \vec{S}_{2,k} + \ldots + w_{3n^2 - 1} \vec{S}_{3n^2 - 1,k} + \delta, \vec{S}_{l,k}) \\
&= w_1 cov(\vec{S}_{1,k}, \vec{S}_{l,k}) + \ldots + w_{3n^2 - 1} cov(\vec{S}_{3n^2 - 1,k}, \vec{S}_{l,k}) \\
&\quad + cov(\vec{\delta}, \vec{S}_{l,k})
\end{aligned}
$$

The JPEG compression is a non-adaptive method that is independent of the pixel values. Therefore, the perturbing term $\vec{\delta}$ is assumed to be independent of the coefficient vector $\vec{S}_{l,k}$ and, consequently, $cov(\vec{\delta}, \vec{S}_{l,k}) = 0$. The covariance reduces the negative impact of the JPEG and double JPEG compression. When l varies from 1 to $3n^2 - 1$, we construct the covariance matrix containing $3n^2 - 1$ linear equations, and the interpolation coefficients w_l are computed as:

$$
\begin{bmatrix} w_1 \\ w_2 \\ \vdots \\ w_{3n^2-1} \end{bmatrix} = \begin{bmatrix} cov(\vec{S}_{1,k}, \vec{S}_{1,k}) & \cdots & cov(\vec{S}_{3n^2-1,k}, \vec{S}_{1,k}) \\ cov(\vec{S}_{1,k}, \vec{S}_{2,k}) & \cdots & cov(\vec{S}_{3n^2-1,k}, \vec{S}_{2,k}) \\ \vdots & \ddots & \vdots \\ cov(\vec{S}_{1,k}, \vec{S}_{3n^2-1,k}) & \cdots & cov(\vec{S}_{3n^2-1,k}, \vec{S}_{3n^2-1,k}) \end{bmatrix}^{-1}
$$
$$
* \begin{bmatrix} cov(\vec{P}', \vec{S}_{1,k}) \\ cov(\vec{P}', \vec{S}_{2,k}) \\ \vdots \\ cov(\vec{P}', \vec{S}_{3n^2-1,k}) \end{bmatrix} \tag{1}
$$

In the interpolation operation, pixels at different interpolated locations usually have different interpolation coefficients. Therefore, it is necessary to obtain the interpolation coefficients separately for the different pixel categories. In the case of the commonly used Bayer pattern, the eight missing color components in a 2×2 Bayer CFA unit are placed in seven categories. The two missing G components are grouped together in one category because of their symmetric interpolation pattern. The remaining six color components are placed in separate categories.

In each category, the interpolation coefficients are assumed to be the same and are computed using Equation (1). The interpolation neighborhood size is $n = 7$ in order to collect more information introduced by the interpolation algorithm while keeping the computation complexity reasonable. Each of the seven categories has $(3 \times 7^2 - 1) = 146$ interpolation

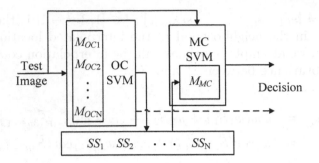

Figure 1.   Combined classification framework.

coefficients. Therefore, the total number of interpolation coefficients is $(3 \times 7^2 - 1) \times 7 = 1,022$.

2.1   Combined Classification Framework

Several researchers have employed multi-class classifiers for camera identification [1, 4, 8, 11, 14]. The methodology involves extracting feature vectors from several image samples created by various camera models. The multi-class classifier is then trained by inputting feature vectors from sample images along with their class labels. After training is complete, the classifier is provided with the feature vector corresponding to a test image; classification is performed by assigning to the test image, the class label corresponding to the class that is the best match. The problem with this approach is that a multi-class classifier cannot identify outliers that do not belong to any of the classes in the original training set.

To address this issue we combine a one-class SVM [13] and a multi-class SVM [2]. The one-class SVM distinguishes outliers that do not correspond to any of the training camera models. If the one-class SVM associates a test image with multiple camera models, the multi-class SVM is used to determine the camera model that is the best match.

Figure 1 presents the combined classification framework. $M_{oc1}, M_{oc2}, \cdots, M_{ocN}$ denote the one-class models and $SS_1, SS_2, \cdots, SS_N$ denote the sets of image samples captured by N cameras. When the feature vector of a test image is extracted, all of the one-class models are first used to classify the test image. Each one-class SVM identifies the image as either being from the camera model it recognizes or an outlier to the model. Each positive result of a one-class SVM indicates that the test image may belong to one of the camera models. In general, there are three possible one-class SVM outputs for a test image:

Table 1. Camera models used in the experiments.

Camera Model	ID	Number of Images
Canon PowerShot A700	1	35
Canon EOS 30D	2	40
Canon PowerShot G5	3	33
Sony DSC-H5	4	35
Nikon E7900	5	35
Kodak Z740	6	35
Kodak DX7590	7	35
Samsung Pro 185	8	35
Olympus Stylus 800	9	35
Fuji FinePix F30	10	37
Fuji FinePix S9500	11	35
Panasonic DMC-FZ8	12	38
Casio EX-Z750	13	35
Minolta Dimage EX 1500	14	44
Canon PowerShot G6	15	37
Olympus E-10	16	31

1. **Outlier:** The test image has been created by an unknown camera outside the data set. In this case, the outlier camera can be exposed by the one-class SVM.

2. **One Positive Result:** The test image has been created by the camera model corresponding to the positive result.

3. **Multiple Positive Results:** The test image has been created by one of the camera models with a positive result.

For Cases 1 and 2, the final decision about image origin is made as indicated by the dashed line in Figure 1. For Case 3, the one-class SVM output is used to select image samples created by the camera models that give positive results for the test image. A new multi-class model M_{MC} is then trained using the selected image samples; this model is used to classify the test image as indicated by the solid line in Figure 1.

3.   Experimental Results

Our experiments used a dataset containing images from sixteen different cameras over 10 brands (Table 1). The images were captured under a variety of uncontrolled conditions, including different image sizes, lighting conditions and compression quality. Each image was divided into 512×512 non-overlapping blocks of which eight were randomly chosen

Table 2.   Confusion matrix for all sixteen cameras.

	1	2	3	4	5	6	7	8	9	10	11	12	13	14	Outlier
1	97.5	*	*	*	*	*	*	*	*	*	*	*	*	*	*
2	1.3	91.9	*	*	*	*	1.3	*	2.5	*	*	*	*	*	*
3	1.9	*	92.3	*	*	*	*	*	*	*	*	*	*	*	*
4	*	*	*	91.7	*	*	*	*	*	*	1.6	*	*	*	*
5	*	*	*	*	90.0	*	*	1.6	*	*	*	*	2.5	*	*
6	*	1.6	*	*	*	90.0	2.5	*	*	*	*	*	*	*	*
7	*	1.6	*	*	*	2.5	90.8	*	*	1.6	*	*	*	*	*
8	*	*	*	*	1.6	*	*	93.3	*	*	*	*	*	*	*
9	*	1.6	*	3.3	*	*	*	*	85.0	2.5	*	3.3	*	*	*
10	*	*	*	*	*	*	*	*	1.5	94.1	2.2	*	*	*	*
11	*	*	*	*	1.6	*	*	*	*	3.3	90.8	*	*	*	*
12	*	*	*	1.4	*	*	*	*	1.4	*	*	93.8	*	*	*
13	1.6	*	*	*	1.6	*	*	*	*	*	*	*	91.7	*	*
14	*	*	*	*	*	*	*	*	*	*	*	*	*	97.9	*
15	10.8	*	11.5	*	1.4	*	*	*	*	*	3.7	*	*	*	72.0
16	*	2.0	*	*	*	*	5.7	*	20.2	*	*	*	1.2	*	69.8

for analysis. The image database consisted of 4,600 different images with 512 × 512 resolution. For each of the first fourteen camera models (IDs 1–14), 160 randomly chosen images were used for classifier training; the remainder were used for testing purposes. The remaining two cameras (IDs 15 and 16) provided 544 images that were used as outlier test cases.

One-class SVM and multi-class SVM implementations provided by LIBSVM [3] were used to construct the classifier. The RBF kernel was used in both SVMs. The kernel parameters were determined by a grid search as suggested in [6].

The experimental results are shown in Table 2 in the form of a confusion matrix. The fifteen columns correspond to the fourteen one-class training models and the outlier class. The sixteen rows correspond to the sixteen cameras used in the study. The (i, j^{th}) element in the confusion matrix gives the percentage of images from camera model i that are classified as belonging to camera model j. The diagonal elements indicate the classification accuracy while elements (15, 15) and (16, 15) indicate the classification performance for the outlier camera models. Values that are less than 1% are denoted by the symbol "*" in the table. The average classification accuracy is 92.2% for the fourteen cameras and 70.9% for the two outlier camera models.

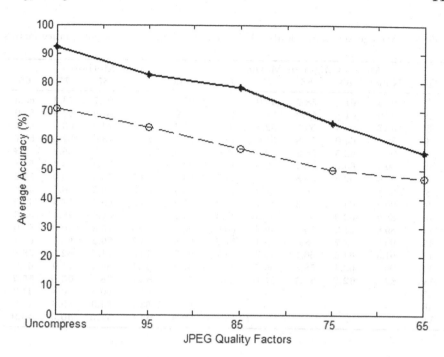

Figure 2. Average accuracy under different JPEG quality factors.

In order to test the classification method on double-compressed JPEG images, the images were re-compressed with secondary quality factors (QF) of {65, 75, 85, 95} without any other manipulations such as scaling and color reduction. Figure 2 presents the performance of the method for various quality factors for the double JPEG compression. The solid line shows the average detection accuracy for the fourteen camera models and the dashed line shows the average detection accuracy for the two outliers. For the quality factor of 95, the classification method provides an average accuracy of 82.5% for the fourteen camera models and 64.6% for the two outliers. For the quality factor of 65, the average accuracy drops to 55.7% for the fourteen cameras and 46.9% for the two outlier cameras.

Table 3 compares the results for the proposed method with those reported for the method of Meng, Kong and You [11]. The Meng-Kong-You method supposedly outperforms other source camera identification methods [11]. However, the results in Table 3 indicate that the Meng-Kong-You method cannot handle double-compressed JPEG images and is incapable of detecting outliers. On the other hand, the

Table 3.   Average accuracy (double JPEG compression) for different quality factors.

	Meng-Kong-You Method					Our Method			
QF	None	95	85	75	65	None	95	85	75	65
1	91.5	61.7	58.7	49.9	50.1	97.5	94.2	91.7	73.3	63.3
2	87.9	61.3	57.2	52.3	47.2	91.9	87.5	81.9	67.5	51.9
3	89.7	60.4	59.5	52.3	50.9	92.3	79.8	74.0	63.5	57.7
4	89.1	64.0	57.8	51.0	49.0	91.7	84.2	78.3	62.5	51.7
5	91.2	62.5	58.8	50.5	47.6	90.0	78.3	73.3	59.2	48.3
6	90.5	65.0	57.4	52.5	49.9	90.0	77.5	75.8	63.3	55.8
7	89.0	62.6	56.8	52.1	48.8	90.8	79.2	75.8	60.0	54.2
8	86.8	62.4	56.2	51.2	48.1	93.3	85.8	79.2	66.7	45.8
9	90.8	64.5	56.3	52.1	49.9	85.0	73.3	69.2	57.5	49.2
10	88.9	62.9	58.7	51.3	49.3	94.1	87.5	81.6	76.5	59.6
11	89.8	61.3	56.8	49.7	50.6	90.8	78.3	75.8	70.0	55.0
12	90.7	63.7	58.0	49.4	51.7	93.8	83.3	79.2	68.1	61.1
13	91.3	64.5	56.1	49.7	48.7	91.7	75.0	71.7	62.5	58.3
14	90.4	60.4	58.8	50.7	51.1	97.9	91.1	88.0	71.9	67.2
Av.	**89.8**	**62.7**	**57.7**	**51.0**	**49.5**	**92.2**	**82.5**	**78.3**	**65.9**	**55.7**
15	-	-	-	-	-	72.0	65.5	60.5	50.3	47.0
16	-	-	-	-	-	69.8	63.7	54.0	49.6	46.8
Av.	-	-	-	-	-	**70.9**	**64.6**	**57.3**	**50.0**	**46.9**

proposed method is robust against double JPEG compression and can detect training model outliers with reasonable accuracy.

4.   Conclusions

This paper has described a new method for determining the source camera for digital images. A covariance matrix is used to obtain a feature vector of 1,022 CFA interpolation coefficients. The feature vector is input to a classifier that is a combination of one-class and multi-class SVMs. The classifier can identify camera models in the training set as well as outliers. Experiments indicate that average accuracies of 92.2% and 70.9% are obtained for camera model identification and outlier camera model identification, respectively. The experiments also demonstrate that the method exhibits good robustness for double-compressed JPEG images.

Acknowledgements

This research was supported by the National High Technology Research and Development Program of China (Program 863; Grant No. 2008AA01Z418) and by the National Natural Science Foundation of China (Grant No. 60572111).

References

[1] S. Bayram, H. Sencar and N. Memon, Improvements on source camera model identification based on CFA interpolation, in *Advances in Digital Forensics II*, M. Olivier and S. Shenoi (Eds.), Springer, Boston, Massachusetts, pp. 289–299, 2006.

[2] B. Boser, I. Guyon and V. Vapnik, A training algorithm for optimal margin classifiers, *Proceedings of the Fifth Annual Workshop on Computational Learning Theory*, pp. 144–152, 1992.

[3] C. Chang and C. Lin, LIBSVM: A Library for Support Vector Machines, Department of Computer Science and Information Engineering, National Taiwan University, Taipei, Taiwan (www.csie .ntu.edu.tw/~cjlin/libsvm).

[4] K. Choi, E. Lam and K. Wong, Automatic source camera identification using intrinsic lens radial distortion, *Optics Express*, vol. 14(24), pp. 11551–11565, 2006.

[5] Z. Geradts, J. Bijhold, M. Kieft, K. Kurosawa, K. Kuroki and N. Saitoh, Methods for identification of images acquired with digital cameras, *Proceedings of the SPIE*, vol. 4232, pp. 505–512, 2001.

[6] C. Hsu, C. Chang and C. Lin, A Practical Guide to Support Vector Classification, Department of Computer Science and Information Engineering, National Taiwan University, Taipei, Taiwan (www .csie.ntu.edu.tw/~cjlin/papers/guide/guide.pdf), 2003.

[7] N. Khanna, A. Mikkilineni, A. Martone, G. Ali, G. Chiu, J. Allebach and E. Delp, A survey of forensic characterization methods for physical devices, *Digital Investigation*, vol. 3(S1), pp. 17–18, 2006.

[8] M. Kharrazi, H. Sencar and N. Memon, Blind source camera identification, *Proceedings of the International Conference on Image Processing*, vol. 1, pp. 709–712, 2004.

[9] Y. Long and Y. Huang, Image based source camera identification using demosaicking, *Proceedings of the Eighth IEEE Workshop on Multimedia Signal Processing*, pp. 419–424, 2006.

[10] J. Lukas, J. Fridrich and M. Goljan, Detecting digital image forgeries using sensor pattern noise, *Proceedings of the SPIE*, vol. 6072, pp. 362–372, 2006.

[11] F. Meng, X. Kong and X. You, A new feature-based method for source camera identification, in *Advances in Digital Forensics IV*, I. Ray and S. Shenoi (Eds.), Springer, Boston, Massachusetts, pp. 207–218, 2008.

[12] T. Ng and S. Chang, Passive-blind image forensics, in *Multimedia Security Technologies for Digital Rights*, W. Zeng, H. Yu and C. Lin (Eds.), Academic Press, New York, pp. 383–412, 2006.

[13] B. Scholkopf, A. Smola, R. Wiliamson and P. Bartlett, New support vector algorithms, *Neural Computation*, vol. 12(5), pp. 1207–1245, 2000.

[14] A. Swaminathan, M. Wu and K. Liu, Component forensics of digital cameras: A non-intrusive approach, *Proceedings of the Fortieth Annual Conference on Information Sciences and System*, pp. 1194–1199, 2006.

[15] A. Swaminathan, M. Wu and K. Liu, Non-intrusive component forensics of visual sensors using output images, *IEEE Transactions on Information Forensics and Security*, vol. 2(1), pp. 91–106, 2007.

Chapter 9

FORENSIC ANALYSIS OF THE SONY PLAYSTATION PORTABLE

Scott Conrad, Carlos Rodriguez, Chris Marberry and Philip Craiger

Abstract   The Sony PlayStation Portable (PSP) is a popular portable gaming device with features such as wireless Internet access and image, music and movie playback. As with most systems built around a processor and storage, the PSP can be used for purposes other than it was originally intended – legal as well as illegal. This paper discusses the features of the PSP browser and suggests best practices for extracting digital evidence.

Keywords: Sony PlayStation Portable, forensic analysis

1.   Introduction

The Sony PlayStation Portable (PSP) is a popular portable video game system that has additional multimedia and Internet-related capabilities. Originally released in 2004, the PSP features a 4.3" widescreen LCD with 480×272 pixel resolution. It comes with a dual core 222 MHz R4000 CPU, 32 MB RAM and 4 MB of embedded DRAM, which holds the operating system [6]. The PSP uses a proprietary Universal Media Disk (UMD) as its primary read-only storage media for games and movies. The device also features 802.11b Wi-Fi connectivity for multiplayer games and utilizes a Pro Duo memory stick for secondary storage.

In September 2007, Sony released a new version of the PSP that is 33% lighter and 19% thinner, appropriately dubbed the PSP Slim & Lite. The Slim & Lite version caches UMD data in memory to decrease game loading time and provides additional features such as a brighter screen, composite TV output, charging via USB and double the onboard RAM (64 MB) [8].

The PSP has updatable firmware that can be downloaded from a Sony server using the Wi-Fi interface. Version 2.0 of the firmware provides a

G. Peterson and S. Shenoi (Eds.): Advances in Digital Forensics V, IFIP AICT 306, pp. 119–129, 2009.

browser and a Really Simple Syndication (RSS) reader. The RSS reader can connect and pull in content via RSS "feeds" typically provided by rapidly updated websites that can be viewed outside of a web browser. The content includes blog entries, news headlines, audio and video. RSS allows for subscriptions to favored content and aggregated feeds from multiple sites [3]. Because these feeds are completely user-defined, they can provide considerable information about the browsing habits of users.

Sony selected the NetFront browser from Access as the internal web browser for the PSP. NetFront is currently deployed in more than 139 devices, including mobile phones, PDAs, digital TVs, gaming consoles and automobile telematics systems from 90 major Internet device manufacturers [10]. The browser has robust capabilities via features such as HTML 4.01 support, flash support, CSS support, tabbed browsing, offline browsing, SSL support, streaming downloads and Smart-Fit rendering [5]. In addition, NetFront provides features associated with traditional web browsers, including the ability to save bookmarks and URL history, both of which provide additional information about the browsing habits of users.

This paper examines the principal features of the PSP browser, in particular, the data structures used to save bookmarks, URL history and other information about user browsing habits. It also presents forensically-sound techniques that can be used to extract digital evidence from the Sony PSP.

2.   Background

In April 2005, a DNS redirection flaw was discovered in the content downloading feature of the *Wipeout Pure* video game that enables web pages other than the official game website to be displayed. This discovery drew attention to the fact that addresses such as `file:///disc0:/` enable UMD files to be viewed; these files are normally hidden from users. Soon after the discovery, a method for formatting PSP executables (e.g., `EBOOT.BIN`) was devised, which brought the inner workings of the PSP to light. (The `EBOOT` file is a packaged executable file, much like a traditional `.exe` file.) Some time later, a hacker named "NEM" and the "Saturn Expedition Committee" were able to successfully reverse engineer the layout of the executable format [9].

These exploits and others enable programmers to modify the firmware directly, allowing unsigned software and third party ("homebrew") applications to be run on the PSP; this is possible because the PSP lacks a mechanism to verify that executables are digitally signed. It is also possible to execute third party applications from a memory stick by modifying Version 1.00 of the Sony firmware. In fact, every version of the PSP firmware has been modified and countless homebrew applications have been developed for the PSP.

Homebrew applications are not always designed for illicit purposes, although some exist solely to circumvent copyright protection. Quite often, they are a way for independent developers to demonstrate their creativity by creating their own PSP games.

3. Memory

The PSP memory stick has a FAT16 file system. Thus, standard forensic software, such as Encase, FTK and hex editors, can be used to analyze the memory stick. The memory stick used in our tests was 1 GB in size; the cluster size in the FAT16 file system was 32 KB.

The PSP also has a significant amount of RAM or cache memory. However, there is no way to directly access the memory and copy the contents other than to physically remove chips from the PSP (which can be extremely risky). Custom firmware is available to obtain a memory dump, but the techniques may not be forensically sound. Also they present a *Catch 22* situation: the only practical way to extract data from PSP RAM is to install software that overwrites some of the RAM data. However, this is not a serious problem because PSP RAM does not hold important user data; it almost exclusively stores firmware and various system settings. In fact, the only user-generated data stored in RAM is the background picture, and only if the user has changed it from the default picture. For this reason, the rest of this paper focuses exclusively on data stored in the memory stick.

4. Browser History Files

The PSP web browser stores the browsing history in various files on the memory stick, each file reflecting a different aspect of the history. Most of the data in the history files is stored as plaintext and is thus easily searched. For example, a search for "http:" will almost always find at least the historyv.dat file and searches of the data within historyv.dat will usually find the other history files (historyi.dat and historys.dat) if they exist on the memory stick.

Testing revealed that the file system usually begins the next history file two clusters below the beginning of the previous history file, i.e., if historyi.dat begins at the relative hex offset of 170000, then historyv.dat begins at the relative offset of 180000. Also, the browser must be shut down gracefully for the history files to be written to the memory stick. This is because the browser does not constantly write to

```
           0  1  2  3  4  5  6  7  8  9  A  B  C  D  E  F  0123456789ABCDEF
00237FF0  00 00 00 00 00 00 00 00 00 00 00 00 00 00 00 00  ................
00238000  56 65 72 2E 30 31 68 74 74 70 3A 2F 2F 77 77 77  Ver.01http://www
00238010  2E 72 65 75 74 65 72 73 2E 63 6F 6D 01 01 68 74  .reuters.com..ht
00238020  74 70 3A 2F 2F 6E 65 77 73 2E 67 6F 6F 67 6C 65  tp://news.google
00238030  2E 63 6F 6D 01 01 68 74 74 70 3A 2F 2F 6E 65 77  .com..http://new
00238040  73 2E 79 61 68 6F 6F 2E 63 6F 6D 01 01 68 74 74  s.yahoo.com..htt
00238050  70 3A 2F 2F 77 77 77 2E 63 62 73 6E 65 77 73 2E  p://www.cbsnews.
00238060  63 6F 6D 01 01 68 74 74 70 3A 2F 2F 77 77 77 2E  com..http://www.
00238070  66 6F 78 6E 65 77 73 2E 63 6F 6D 01 01 68 74 74  foxnews.com..htt
00238080  70 3A 2F 2F 77 77 77 2E 6E 65 77 73 2E 63 6F 6D  p://www.news.com
00238090  01 01 68 74 74 70 3A 2F 2F 6E 65 77 73 2E 62 62  ..http://news.bb
002380A0  63 2E 63 6F 6F 2E 75 6B 01 01 68 74 74 70 3A 2F  c.co.uk..http://
002380B0  77 77 77 2E 6D 73 6E 62 63 2E 63 6F 6D 01 01 68  www.msnbc.com..h
002380C0  74 74 70 3A 2F 2F 77 77 77 2E 63 6E 6E 2E 63 6F  ttp://www.cnn.co
002380D0  6D 01 01 68 74 74 70 3A 2F 2F 77 77 77 2E 66 61  m..http://www.fa
002380E0  72 6B 2E 63 6F 6D 01 01 68 74 74 70 3A 2F 2F 77  rk.com..http://w
002380F0  77 77 2E 67 6F 6F 67 6C 65 2E 63 6F 6D 01 01 00  ww.google.com...
00238100  00 00 00 00 00 00 00 00 00 00 00 00 00 00 00 00  ................
00238110  00 00 00 00 00 00 00 00 00 00 00 00 00 00 00 00  ................
```

Figure 1. Format of `historyi.dat` pages.

the memory stick while it is being used. Instead, the browser keeps everything in internal memory and writes the history files to the memory stick just before it closes. This means that if the PSP is turned off, or if the memory stick is removed before the web browser is exited, or even if the PSP is returned to the home page without closing the web browser, then the history files are not written to the memory stick.

The first history file, which stores all the manually-typed web addresses, is found in the following location on the memory stick:

X:\PSP\SYSTEM\BROWSER\historyi.dat

Note that X: is the drive letter assigned to the memory stick.

A sample page in `historyi.dat` is shown in Figure 1. The pages in the file have the format:

```
<version number>[typed address](white space)
                 [typed address](white space)
                 [typed address](...)
```

where the **version number** is usually "Ver.01."

The web addresses stored in `historyi.dat` are not necessarily those that were visited; they are merely those that the user manually typed into the browser and attempted to visit. Entries in the file appear exactly as they were typed, i.e., http:// is not automatically added to the beginning of a typed URL. Additionally, the most recently typed addresses are placed at the beginning of `historyi.dat` instead of being appended to the end, and an entry only appears once in the file regardless of how many times it was typed into the browser. When an entry is repeated, it is simply moved to the beginning of the file.

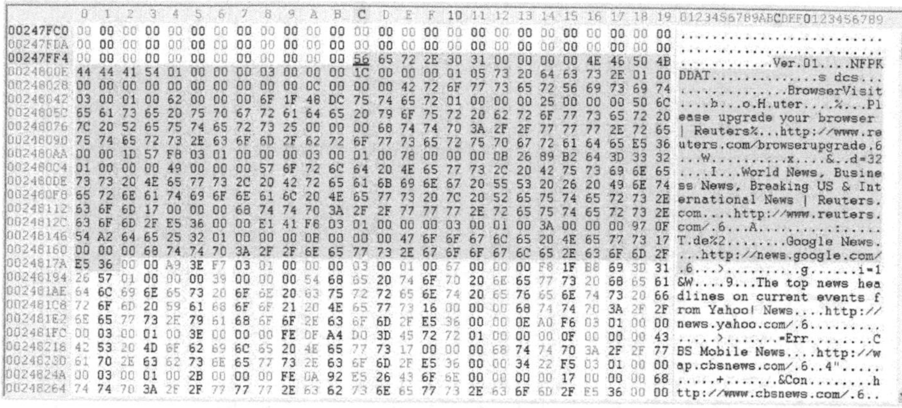

Figure 2. Format of `historyv.dat` entries.

The second history file stores web addresses that are actually visited, whether they are manually typed or accessed via html links:

X:\PSP\SYSTEM\BROWSER\historyv.dat

As with the `historyi.dat` file, the most recently visited web address appears at the beginning of the `historyv.dat` file. Unlike the `historyi.dat` file, entries can appear multiple times if they were accessed more than once. The URLs in `historyv.dat` are always valid web addresses. Also, the `historyv.dat` file can be transferred to the \Browser folder in any PSP memory stick and the PSP browser may be used to display the title, address and last accessed dates of all the entries in the file. This is very useful because the last accessed date is normally stored within the encoded data.

Figure 2 shows a portion of the `historyv.dat` file. The format of each entry in the file is:

```
<version number><encoded data>
      [website HTML title (if applicable)]
      [(URL protocol)(website address)]
      <_6><encoded data>
            [Website HTML title (if applicable)]
            [(URL protocol)(website address)]
      <_6><encoded data>(...)
```

where the `version number` is usually "Ver.01."

5. Internet Search Feature

Sony released Version 4.0 of the PSP firmware in June 2008. This firmware update enables users to perform Internet searches directly from

```
         0  1  2  3  4  5  6  7  8  9  A  B  C  D  E  F  0123456789ABCDEF
001BFFE0 00 00 00 00 00 00 00 00 00 00 00 00 00 00 00 00 ................
001BFFF0 00 00 00 00 00 00 00 00 00 00 00 00 00 00 00 00 ................
001C0000 3C 6C 69 73 74 3E 3C 64 61 74 61 3E 3C 65 6E 67 <list><data><eng
001C0010 69 6E 65 5F 69 64 78 3E 30 3C 2F 65 6E 67 69 6E ine_idx>0</engin
001C0020 65 5F 69 64 78 3E 3C 74 69 74 6C 65 3E 62 62 63 e_idx><title>bbc
001C0030 2B 6E 65 77 73 3C 2F 74 69 74 6C 65 3E 3C 2F 64 +news</title></d
001C0040 61 74 61 3E 3C 64 61 74 61 3E 3C 65 6E 67 69 6E ata><data><engin
001C0050 65 5F 69 64 78 3E 30 3C 2F 65 6E 67 69 6E 65 5F e_idx>0</engine_
001C0060 69 64 78 3E 3C 74 69 74 6C 65 3E 61 62 63 2B 6E idx><title>abc+n
001C0070 65 77 73 3C 2F 74 69 74 6C 65 3E 3C 2F 64 61 74 ews</title></dat
001C0080 61 3E 3C 64 61 74 61 3E 3C 65 6E 67 69 6E 65 5F a><data><engine_
001C0090 69 64 78 3E 30 3C 2F 65 6E 67 69 6E 65 5F 69 64 idx>0</engine_id
001C00A0 78 3E 3C 74 69 74 6C 65 3E 66 61 72 6B 3C 2F 74 x><title>fark</t
001C00B0 69 74 6C 65 3E 3C 2F 64 61 74 61 3E 3C 64 61 74 itle></data><dat
001C00C0 61 3E 3C 65 6E 67 69 6E 65 5F 69 64 78 3E 30 3C a><engine_idx>0<
001C00D0 2F 65 6E 67 69 6E 65 5F 69 64 78 3E 3C 74 69 74 /engine_idx><tit
001C00E0 6C 65 3E 63 6E 6E 3C 2F 74 69 74 6C 65 3E 3C 2F le>cnn</title></
001C00F0 64 61 74 61 3E 3C 2F 6C 69 73 74 3E 00 00 00 00 data></list>....
001C0100 00 00 00 00 00 00 00 00 00 00 00 00 00 00 00 00 ................
001C0110 00 00 00 00 00 00 00 00 00 00 00 00 00 00 00 00 ................
```

Figure 3. Format of `historys.dat` entries.

the PSP Home Menu [11]. It appears that Google is the default search engine.

The `historys.dat` file that stores the corresponding information is located at.

    X:\PSP\SYSTEM\BROWSER\historys.dat

Figure 3 shows a portion of the `historys.dat` file. The format of file entries is:

```
<list><data><engine_idx>[generated number]
        </engine_idx><title>
            [query]</title></data><data>
        <engine_idx>[generated number]
        </engine_idx><title>
            [query]</title></data>...</list>
```

Note that the recording format is very similar to that of common markup languages such as HTML and XML. Unlike the `historyi.dat` and `historyv.dat` files, every time a new query is performed using the Internet Search feature, a new `historys.dat` file is created that shows the new query with the previous queries appended at the end.

6.   RSS History

Version 2.6 of the PSP firmware (released in November 2005) added support for RSS feeds [11]. The RSS Channel feature is presented to users above the Web Browsing option. This mobile RSS aggregator was originally designed for downloading web feeds and pod casts in MP3 or

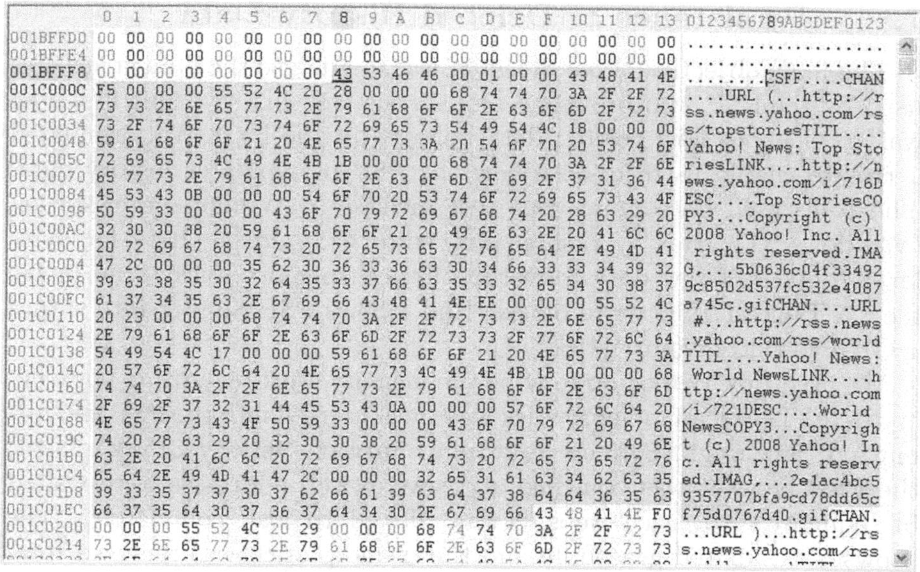

Figure 4. Format of CHLIST entries.

AAC formats. Version 2.8 added support for downloading video and image content [7]. All RSS content may be downloaded directly to a memory stick in the PSP. The data downloaded from a RSS feed is stored in the CHLIST file:

X:\PSP\SYSTEM\RSSCH\CHLIST

Figure 4 shows a portion of the CHLIST file. The format of file entries is:

```
CSFF<binary data>
CHAN<binary data>
URL<binary data>[URL of RSS feed (with http://)]
TITL<binary data>[Title of website]
LINK<binary data>[URL of website behind RSS feed]
DESC<binary data>[Description of website]
COPY<binary data>[Copyright info]
IMAG<binary data>[Name of associated image
                 (also saved in ...\RSSCH]
CHAN<binary data>[etc.]
```

For each entry in the CHLIST file, there is a corresponding image linked to that entry in the same folder (...\RSSCH). Note that the topmost entries in the file are the oldest accessed RSS feeds and the entries towards the end of the file are the most recently accessed RSS feeds.

7.   Persistence of Deleted History Data

The main objective of our research was to analyze the data structures used by the PSP web browser when storing web history. Our first step was to examine how deleted data behaved in the PSP. We formulated a test to enable us to discern how the PSP manipulates history data. The tools used to conduct this test included a Windows workstation (Windows XP), Hex Workshop (hex editor) [1], dd [4] (used for byte-level copying of raw data from the physical memory stick to an image file), a Tableau USB write blocker and a Sony PSP Slim & Lite with a 1 GB Pro Duo memory stick.

We used Hex Workshop to wipe the memory stick by writing 00 to every byte. The clean memory stick was then inserted into the PSP and formatted (Settings>System Settings>Format>Memory Stick). Next, the PSP browser was launched and several web addresses were visited in sequence after each website was allowed to load completely. The browser was exited gracefully and the memory stick was removed from the PSP and connected to a write blocker. A raw image of the memory stick was created using dd and saved as Before.001. The memory stick was then removed from the write blocker and connected to the workstation. The history files were manually deleted from the directory X:\PSP\SYSTEM\BROWSER using Windows Explorer. The memory stick was placed back in the PSP and several new web addresses were accessed via the PSP browser. The memory stick was then removed from the PSP and connected back to the write blocker and another raw image was created (After1.001). Next, dd was used to restore the image Before.001 to the memory stick. The memory stick was placed back in the PSP, the browser was launched and the history was cleared completely using the following steps:

```
History>Options>Delete All
Tools>Delete Cookies
Tools>Delete Cache
Tools>Delete Authentication
Tools>Delete Input History
```

After the history was deleted, a second set of web addresses was visited using the browser. The memory stick was removed from the PSP, connected again to the write blocker and a raw image was created (After2.001). The three images, Before.001, After1.001 and After2.001, were compared and analyzed using a hex editor.

After analyzing the files, it was discovered that when the history is erased from the PSP browser using the method mentioned above (image After2.001), the browser generally does not overwrite the old history when it begins a new history. Instead, the file system (most of the time) simply moves down one cluster from the beginning of the old history file. In another words, if the old `historyi.dat` begins at relative hex offset 170000, then the new `historyi.dat` begins at 178000, which means that both the old `historyi.dat` and the old `historyv.dat` (which are usually located at an offset of 180000) would not be overwritten.

In contrast, when the browser history is deleted using Windows Explorer (`After1.001`), the old history is generally overwritten by the new history. This is because the PSP does not move down a cluster before saving the new data. However, if the new history is smaller than the old history (i.e., if the old history has twenty entries and the new history only has ten entries), then parts of the old history are still recoverable.

We discovered that the only data that was consistently altered was the data located in the FAT. However, the first few bytes of each deleted history were almost always changed to indicate that they were deleted and not active. Depending on the circumstances, the actual history files were untouched, not entirely overwritten or completely unrecoverable. In general, however, the closer the history files are to the end of the memory stick, the longer they survive.

8.   Persistence of Overwritten Data

Peculiar behavior was observed when performing the test described above to study the persistence of deleted history data. For some reason, overwritten data in the hex address range of 168000-1FFFFF can be recovered completely by formatting the memory stick. This behavior was confirmed by connecting a memory stick directly to the workstation and completely filling it with aa values using a hex editor. After it was confirmed that the memory stick only had aa values written to it, it was completely filled again, but this time with bb values. The memory stick was then placed in the PSP and formatted using the built-in function. Finally, the memory stick was connected to a write blocker and a search for aa values was conducted; these were always the only values stored in the 168000-1FFFFF address range. The rest of the memory stick was filled with bb values, except for the locations holding the file system. This test was conducted several times with values other than aa and bb. One test even used the 16-byte pattern 12 23 34 45 56 67 78 89 90 0a ab bc cd de ef f1 instead of aa. Nevertheless, the overwritten data always reappeared after being formatted.

This peculiar behavior does not appear to be due to some unknown function of the PSP; rather, it is due to the combination of some physical property of the memory stick and the way the format function works.

This was verified by placing a memory stick in a PSP, writing history files to it and then having the new history files overwrite the old files. The memory stick was then formatted on a completely different PSP and the same behavior was observed. Since it was not possible for the second PSP to rewrite the history files that were created by first PSP, it is apparent that the behavior is not caused by the PSP directly; instead, it has something to do with a physical property of the memory stick. Finally, the fact that the behavior was not observed when the memory stick was formatted by the Windows workstation shows that the PSP formatting function is involved rather than a specific PSP device.

9.   Conclusions

The Sony PSP is not merely a portable gaming console, but a sophisticated device with considerable storage capacity and Internet access. Indeed, it provides the functionality of a small personal computer.

Our research demonstrates that it is possible to recover web browsing history and RSS subscription information from a PSP. Several methods have been proposed for identifying and recovering this information. But further research is required to examine the forensic implications of other PSP features, especially as Sony continues to develop PSPs with new functionality.

Digital forensic investigators are certain to encounter increasing numbers of Sony PSPs and other gaming devices in their crime scene investigations. A modified Xbox [2] is capable of running Linux applications; other game devices can run Linux without any modifications. Consequently, it is important that digital forensic researchers focus on gaming devices, conduct comprehensive examinations of their advanced features, determine the locations where evidence may reside, and develop forensically-sound methodologies for recovering the evidence.

References

[1] BreakPoint Software, Hex Workshop, Cambridge, Massachusetts (www.hexworkshop.com).

[2] P. Burke and P. Craiger, Forensic analysis of Xbox consoles, in *Advances in Digital Forensics III*, P. Craiger and S. Shenoi (Eds.), Springer, Boston, Massachusetts, pp. 269–280, 2007.

[3] R. Cadenhead, G. Smith, J. Hanna and B. Kearney, The application/rss+xml media type, Network Working Group (www.rssboard .org/rss-mime-type-application.txt), 2006.

[4] Free Software Foundation, dd: Convert and copy file, GNU Core-utils, Boston, Massachusetts (www.gnu.org/software/coreutils/man ual/html_node/dd-invocation.html).

[5] J. Puente, What browser does the Sony PSP use? (jefte.net/psp /what-browser-does-the-sony-psp-use), 2006.

[6] J. Sanches, PSP Slim & Lite, Steel Media, Uxbridge, United Kingdom (www.pocketgamer.co.uk/r/PSP/PSP+Slim+&+Lite/hardw are_review.asp?c=4188), September 18, 2007.

[7] Sony Computer Entertainment America, RSS document speci-fications, Culver City, California (www.playstation.com/manual /psp/rss/en/spec.html), 2007.

[8] Sony Computer Entertainment America, Sony Computer Enter-tainment America to offer limited-edition entertainment packs with newly designed PSP (PlayStation Portable) starting this fall, Culver City, California (www.us.playstation.com/News/PressRel eases/407), July 11, 2007.

[9] Wikipedia, PlayStation Portable homebrew, Wikipedia Foundation, San Francisco, California (en.wikipedia.org/wiki/Psp_homebrew).

[10] WindowsForDevices.com, Access NetFront browser wins best soft-ware award, New York (www.windowsfordevices.com/news/NS791 1853350.html), November 8, 2004.

[11] Xtreme PSP, Firmware history and compatibility (www.xtreme psp.com/firmware.php).

III

INTEGRITY AND PRIVACY

Chapter 10

IMPLEMENTING BOOT CONTROL FOR WINDOWS VISTA

Yuki Ashino, Keisuke Fujita, Maiko Furusawa, Tetsutaro Uehara and Ryoichi Sasaki

Abstract   A digital forensic logging system must prevent the booting of unauthorized programs and the modification of evidence. Our previous research developed Dig-Force2, a boot control system for Windows XP platforms that employs API hooking and a trusted platform module. However, Dig-Force2 cannot be used for Windows Vista systems because the hooked API cannot monitor booting programs in user accounts. This paper describes an enhanced version of Dig-Force2, which uses a TPM and a white list to provide boot control functionality for Windows Vista systems. In addition, the paper presents the results of security and performance evaluations of the boot control system.

Keywords: Evidence integrity, boot control, Windows Vista

1.   Introduction

Personal computers are often the instruments and/or victims of electronic crime. This makes it important to securely log and store all potential evidence for use in legal proceedings [5]. The logging system should operate in a "sterile" environment, log and store all operational data and enable a third party to verify the integrity of the logged data.

We previously designed the Dig-Force system [1] to address these issues. Dig-Force reliably records data pertaining to computer usage on the computer itself and uses chained signatures to maintain the integrity of the evidentiary data. Dig-Force has been shown to be effective even on standalone computers located outside a protected network. Maintaining the security of Dig-Force requires an environment that prevents the execution of boot jamming programs as well as the modification of the Dig-Force program itself.

G. Peterson and S. Shenoi (Eds.): Advances in Digital Forensics V, IFIP AICT 306, pp. 133–141, 2009.

Our next version, Dig-Force2 [2], was developed to maintain a secure environment under Windows XP. Dig-Force2 implements boot control using API hooking and incorporates a trusted platform module (TPM) [7] to prevent boot jamming programs from executing and to detect modifications to Dig-Force. Dig-Force2 runs as a Windows service and hooks the `RtlCreateProcessParameters` API [3]. It verifies that any booting program that executes is non-malicious using a white list and TPM. Only a booting program on the white list is allowed to execute.

Unfortunately, Dig-Force2 does not operate on Windows Vista because the `RtlCreateProcessParameters` API hook cannot be used to monitor booting programs in user accounts. This paper describes an enhanced version of the boot control system, named Boot Control Function for Windows Vista (BCF/Vista). The enhanced system uses a TPM and white list with a controller process that runs as a Windows service, along with an agent process that executes within the user account.

2.   Dig-Force2 Boot Control

Dig-Force2 [2] is designed to be used by administrators, users and system verifiers. Administrators are responsible for setting up and configuring computer systems. Users operate computers using their assigned Windows XP accounts, which are referred to as "user accounts." System verifiers are responsible for verifying the log files created by the system. An administrator can also serve as a system verifier.

Figure 1 presents the Dig-Force2 architecture. It has five components: (i) logging module, (ii) storage module, (iii) boot control function, (iv) Windows service, and (v) user account. The logging module captures data about computer operations (e.g., user actions and system behavior) and sends it to the storage module. The storage module tags the received data with the date, time and data type. Additionally, it digitally signs the data with a hysteresis signature [4, 6] using a public key stored in the TPM before writing the data to the log file. The hysteresis signature ensures that any alterations to the data in the log file are detected [1].

It is essential that boot jamming programs are prevented from executing and that the logging and storage modules are not modified in an unauthorized manner. In order to maintain such a secure environment, Dig-Force2 hooks the `RtlCreateProcessParameters` API, which is invoked whenever a Windows XP program starts. Dig-Force2 then checks that the program is on the white list before permitting it to execute. The check involves computing a hash value of the program and comparing the value with the matching digital signature from the white list after decrypting it using a public key stored in the TPM.

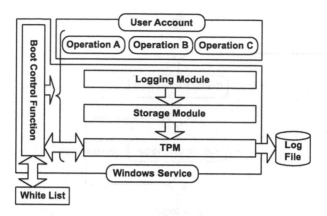

Figure 1. Dig-Force2 architecture.

3. Boot Control Function for Windows Vista

Dig-Force2 runs as a Windows service under Windows XP, which means that it cannot be stopped by anyone except the administrator. However, in Windows Vista, the hooking program must run under a user account if Dig-Force2 is to hook programs executing in a user account. This means that the user can stop the hooking program at any time using the Windows Task Manager.

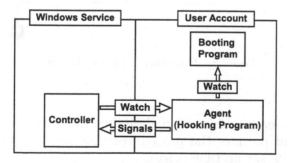

Figure 2. Boot Control Function (Windows Vista) architecture.

Figure 2 presents the architecture of the Boot Control Function for Windows Vista (BCF/Vista). The agent is an enhanced hooking program that executes in the user account; it monitors all booting programs executed from the user account and communicates their condition to the controller. The controller, which is implemented as a Windows service, prevents a user from terminating the agent.

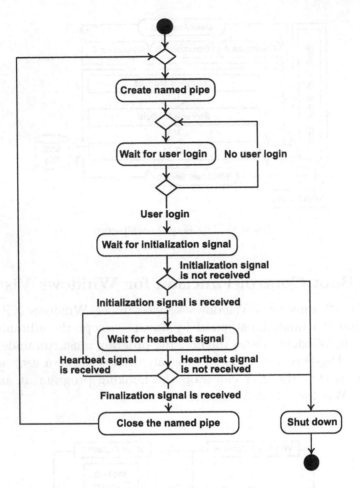

Figure 3. Controller logic flow diagram.

The remainder of this section describes the logic of the controller and agent, and outlines the procedures that must be followed by an administrator to set up BCF/Vista.

3.1   Controller

The controller is an administrator-level Windows service, which ensures that the agent hooking program is always running. Figure 3 presents the controller logic, which involves four main steps. First, the controller creates a named pipe to communicate with an agent executing in a user account. It then waits for an initialization signal from the agent which indicates that a user has logged on. If an initialization signal from

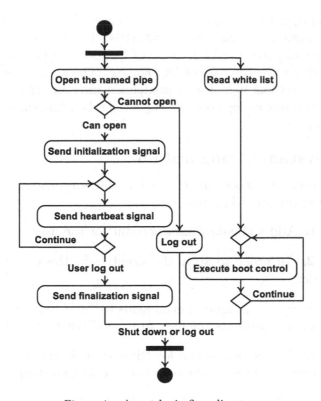

Figure 4. Agent logic flow diagram.

the agent is not received by the controller within a specified time interval, the controller shuts down Vista. The controller then listens for a heartbeat signal from the agent, which indicates that the agent is executing. If the heartbeat signal is not received by the controller, Vista is shut down. When the user logs off, the controller receives a finalization signal from the agent, upon which it closes the named pipe.

3.2 Agent

The agent operates in a manner similar to Dig-Force2 except that the `CreateProcessW` and `CreateProcessA` APIs are hooked instead of `RtlCreateProcessParameters`. However, the BCF/Vista agent runs in a user account to enable boot process hooking unlike Dig-Force2, which runs as a Windows service.

Figure 4 presents the boot control logic of the agent. The agent first opens the named pipe created by the controller. If the agent cannot open the pipe, the agent forcibly logs out the user. Next, the agent sends an initialization signal to the controller, reads the white list and hooks the `CreateProcessW` and `CreateProcessA` APIs. This becomes a secondary process that compares the hash value of the booting program with the white list value as in the case of Dig-Force2. The agent then sends the controller a heartbeat signal at a predefined interval. If the user shuts down the computer or logs out, the agent sends a finalization signal to the controller.

3.3   System Configuration

Before a user can access the computer, the administrator must configure it using the following procedure:

- **Step 1:** Add a standard user account for the user.

- **Step 2:** In order to start the agent, add the task to the task scheduler.

- **Step 3:** Set the program permissions to "read only" from the user account; this prevents the user from modifying programs.

- **Step 4:** Set the permission for "Start-Up" in the user account to "read only" to prevent the user from adding start-up programs.

- **Step 5:** Check "Audit Process Tracking" in the local security policy.

- **Step 6:** Install the white list.

- **Step 7:** Add the controller as a Windows service.

- **Step 8:** Enable BitLocker.

- **Step 9:** Set the BIOS password.

- **Step 10:** In the safe mode, enable the booting service (Step 7).

The administrator then sets up the task scheduler as follows:

- **Step 1:** Create a new task schedule in the folder Task Scheduler Library of the task scheduler.

- **Step 2:** Set "When running the task, use the following user account:" to point to the user account.

- **Step 3:** Set "At log on" and "On an event" with target log to "Security" when triggered.

- **Step 4:** Set "Start a program" to the status of "Action" and register the program path of the agent.

- **Step 5:** Uncheck the checkbox of "Start the task only if the computer is on AC power."

4. System Evaluation

We consider four attacks where the attacker is a user with a standard user account: (i) BCF/Vista removal or modification, (ii) white list modification, (iii) BCF/Vista start-up blocking, and (iv) jamming.

In order to remove or modify the BCF/Vista program file, an attacker must have administrator privileges, which is not possible without the administrator's password. Alternatively, the attacker could attempt to access the hard drive directly, for example, by booting another OS from an alternative storage media or installing the hard drive in another machine. However, the BIOS of the computer is set to boot only from the hard drive, which is fully encrypted using BitLocker. These security measures prevent access to the BCF/Vista program files.

The goal of an attack on the white list is to add a program to the list. To do this, the attacker must be able to add the digital signature of the program to the white list. This requires the secret key used to create the list or the creation of a fabricated white list with a fake key pair. But these will not work because the attacker neither has the secret key nor the password required to install a fake public key in the TPM.

The third attack, blocking the booting of BCF/Vista, is not possible because a Windows service cannot be modified without the administrator's password. This password is not available to the user.

A boot jamming attack is effective only if the jamming program starts before `CreateProcessW` and `CreateProcessA` are hooked. Since the agent and the hooking start automatically at login, the attacker would have to add the jamming program to the start-up list. This is not possible because the start-up directory permission level is set to "read only."

The evaluation of BCF/Vista was conducted using a Dell VOSTRO 1310 with a 2.5 GHz Intel Core2 Duo T9300 CPU and 4 GB RAM running Windows Vista Ultimate Edition. The program required 1,080 seconds to create a white list by calculating the hash values and the digital signatures for the 3,273 `.exe` files residing on the machine. The signing process clearly requires a considerable amount of time. However, it is performed only once when the white list is created and, therefore, does not impact normal computer operations.

Table 1. Booting times.

	Trial 1	Trial 2	Trial 3	Trial 4	Trial 5
`notepad.exe`	1.5491	0.0105	0.0104	0.0105	0.0104
`calc.exe`	1.4850	0.0124	0.0126	0.0124	0.0125
Microsoft Word 2007	1.6446	0.0129	0.0127	0.0132	0.0127
Microsoft PowerPoint 2007	1.5849	0.0166	0.0169	0.0165	0.0169

Table 1 shows the time taken to boot four programs, where each program was booted five times in succession (Trials 1–5). The boot period of a program is measured from the time the boot control function permits the program to boot to the time when `CreateProcessW` and `CreateProcessA` are called.

Note that a significant amount of time (1.4850 to 1.6446 seconds) is required for booting during the first trial when BCF/Vista has to read the white list. After this, the booting time is much less because BCF/Vista uses the white list data, which was read the first time it was executed. The booting time is, thus, acceptable and BCF/Vista has negligible impact on program operation.

5.   Conclusions

BCF/Vista provides a secure and reliable environment for logging data pertaining to computer operations. In particular, it preserves the integrity of evidence by preventing the booting of unauthorized programs and evidence modification. The architecture, which uses special controller and agent processes, a TPM and a white list to provide boot control functionality for Windows Vista systems, has a negligible impact on system performance.

References

[1] Y. Ashino and R. Sasaki, Proposal of digital forensic system using security device and hysteresis signature, *Proceedings of the Third International Conference on Intelligent Information Hiding and Multimedia Signal Processing*, pp. 3–7, 2007.

[2] K. Fujita, Y. Ashino, T. Uehara and R. Sasaki, Using boot control to preserve the integrity of evidence, in *Advances in Digital Forensics IV*, I. Ray and S. Shenoi (Eds.), Springer, Boston, Massachusetts, pp. 61–74, 2008.

[3] Microsoft Corporation, Services, Redmond, Washington (msdn .microsoft.com/en-us/library/ms685141.aspx).

[4] K. Miyazaki, S. Susaki, M. Iwamura, T. Matsumoto, R. Sasaki and H. Yoshiura, Digital document sanitizing problem, *Institute of Electronics, Information and Communication Engineers Technical Reports*, vol. 103(195), pp. 61–67, 2003.

[5] R. Sasaki, Y. Ashino and T. Masubuchi, A trial for systematization of digital forensics and proposal on the required technologies, *Japanese Society of Security Management Magazine*, April 2006.

[6] S. Susaki and T. Matsumoto, Alibi establishment for electronic signatures, *Transactions of the Information Processing Society of Japan*, vol. 43(8), pp. 2381–2393, 2008.

[7] Trusted Computing Group, Beaverton, Oregon (www.trustedcom putinggroup.org).

Chapter 11

A FORENSIC FRAMEWORK FOR HANDLING INFORMATION PRIVACY INCIDENTS

Kamil Reddy and Hein Venter

Abstract   This paper presents a framework designed to assist enterprises in implementing a forensic readiness capability for information privacy incidents. In particular, the framework provides guidance for specifying high-level policies, business processes and organizational functions, and for determining the device-level forensic procedures, standards and processes required to handle information privacy incidents.

Keywords: Forensic readiness capability, information privacy incidents

1.   Introduction

Information privacy is the interest individuals have in accessing, controlling or influencing the use of their personal information [7]. The protection of information privacy is mandated by law in many countries [16, 20]. Enterprises operating in these countries have a legal obligation to secure the information they use. Over and above the legal obligations, consumers [11] and corporate governance standards [12] demand that information privacy be protected regardless of the geographical location of an enterprise.

Digital forensic readiness is a corporate goal involving technical and non-technical actions that maximize the ability of an enterprise to use digital evidence [19]. It ensures the best possible response to incidents that may occur in an enterprise network. Maintaining an effective forensic readiness capability requires carefully considered and coordinated participation by individuals and departments throughout the enterprise [19]. A forensic readiness capability developed or executed in an *ad hoc* manner is unlikely to succeed [8].

G. Peterson and S. Shenoi (Eds.): Advances in Digital Forensics V, IFIP AICT 306, pp. 143–155, 2009.
© IFIP International Federation for Information Processing 2009

The concepts of information privacy and forensic readiness intersect when an information privacy violation occurs and it is necessary to conduct a forensic investigation of the violation. While privacy violations are often the result of security breaches (e.g., unauthorized access to private information), they also occur when private information is used inappropriately by individuals who are authorized to access the information. Therefore, enterprises with a forensic readiness capability for dealing with security-related incidents may not be in an optimal position to respond to privacy-related incidents. To address this issue, we propose a framework that considers the requirements for ensuring forensic readiness with respect to information privacy incidents.

The framework is a theoretical representation of a generic forensic readiness capability for dealing with information privacy violations in an enterprise. As such, it aims to provide a basis upon which enterprises can build a forensic readiness capability for information privacy incidents. Since forensic readiness requires the participation of individuals at all levels and across departmental boundaries [19], the purpose of the framework is to provide guidance at a high level by specifying the appropriate policies, business processes and organizational functions. It also enables an enterprise to determine the device-level forensic procedures, standards and processes required to implement a forensic readiness capability for information privacy incidents.

It is important to note that this paper focuses on the structural aspects of the framework rather than its procedural aspects. Structural aspects refer to the choice of the elements contained in the framework and the relationships between the elements. On the other hand, the procedural aspects merely deal with the practical measures necessary to implement the framework in an enterprise. To our knowledge, little, if any, research focusing on the structural aspects of a forensic readiness framework for handling information privacy incidents has been published.

2.   Related Work

This section discusses related work on forensic readiness and the role of information privacy in digital forensics. It also discusses the "Fair Information Principles" [9], which are at the core of most approaches for protecting information privacy.

The work of Endicott-Popovsky, *et al.* [8] focuses on forensic readiness at the enterprise level. It deals with network forensic readiness as a means for breaking the cycle of attack and defense. Our work is different in that it also addresses information privacy and includes a wider variety of information technologies and business processes.

Other efforts related to forensic readiness have concentrated on tools and techniques [8]. Several researchers have focused on the organizational aspects of forensic readiness. Yasinsac and Manzano [23] have defined policies for computer and network forensics; Wolfe [23] has discussed forensic policies in organizations; Rowlingson [19] has specified a ten step process for implementing forensic readiness; Luoma [13] has proposed the establishment of a multi-disciplinary management team to ensure legal compliance with discovery requests; and Taylor, *et al.* [21] have studied forensic policy specification and its use in forensic readiness.

The vast majority of work related to privacy in the digital forensic literature focuses on protecting the privacy of computer users during forensic investigations [1, 2, 4]. Unfortunately, a comprehensive treatment of information privacy and its impact on forensic readiness has not been conducted.

The Fair Information Principles are a guide for the ethical handling of private information and form the basis for information privacy laws in countries around the world [9]. The eight principles, as espoused by the Organization for Economic Cooperation and Development [17], are: collection limitation, data quality, purpose specification, use limitation, security safeguards, openness, individual participation and accountability. The principles provide a practical definition of information privacy and specify obligations for enterprises with regard to the ethical handling of private information. In addition to covering information privacy, the obligations focus on protecting the confidentiality of data subjects. Enterprises that fail to meet these obligations are likely to be in violation of information privacy laws.

3.   Rationale

Information security has traditionally been concerned with the confidentiality, integrity and availability of information [21]. Information privacy, on the other hand, focuses on the ethical and legal use of information [3]. Confidentiality, integrity and availability are necessary – but insufficient – conditions for information privacy [3]. Thus, information privacy has a wider range of potential violations and incidents since the ethical and legal use requirements are in addition to the traditional requirements for security.

Ethical or legal usage requirements related to information privacy directly affect enterprise business processes. Businesses processes do not specify the boundaries for acceptable use. Ideally, acceptable use is specified via policies [21] derived from authoritative sources such as information privacy laws and ethical guidelines. In some instances, ethical guidelines (such as the Fair Information Principles) may require the creation of "privacy-specific business processes" that deal with private information. An example is a business process that handles requests to access information.

Information technology underlies privacy-related and privacy-specific business processes. In an enterprise, information technology facilitates the execution of business processes that operate on private information. The particular information technologies used in a business process determine to a large extent what can be done with private information. For example, using a database instead of flat text files, makes it easier to query the stored data. Therefore, policies are required to govern the use and configuration of information technologies to ensure that they are used appropriately.

Digital forensic investigations of information privacy incidents in an enterprise involve the information privacy context: privacy-related business processes, privacy-specific business processes, information technologies supporting the processes, policies that govern the processes, and the auditing and monitoring of processes. The information privacy context, with the exception of information technology, expresses what is required by a privacy-specific approach for digital forensic readiness in addition to the traditional security-related approach.

There are two cases in which a forensic readiness capability for information privacy incidents is particularly useful. The first occurs when an entity outside the enterprise violates a subject's information privacy; this situation closely parallels the common security-related scenario of an outsider attacking the enterprise. The second case is internal in nature. An example is when a data subject alleges that the enterprise itself is responsible for the information privacy violation. If the data subject takes legal action, a forensic readiness capability for information privacy incidents would enable the enterprise to conduct an effective digital forensic investigation that can be used in its defense. Another example is when an employee is charged with violating the enterprise's privacy policy. The enterprise may conduct a digital forensic investigation to present evidence against the employee in a disciplinary hearing. The investigation is likely to proceed very efficiently if the enterprise has a mature forensic readiness capability for information privacy incidents.

4.   Forensic Framework

This section describes the framework intended to provide enterprises with a generic forensic readiness capability for dealing with information

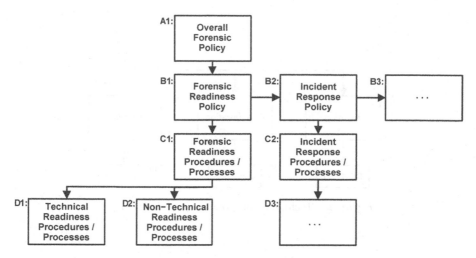

Figure 1. Forensic framework (Levels A – D).

privacy incidents. Due to the size of the framework, we only examine the components that are relevant to handling information privacy incidents.

The forensic framework has a hierarchical tree-like structure with several levels (Figure 1). Each level has various elements depicted as blocks. The blocks within a level (e.g., Level B) are labeled sequentially from left to right (e.g., Blocks B1, B2 and B3).

4.1   Top Levels

At the top of the framework is Block A1, which corresponds to an overall forensic policy that has been approved by management. The forensic policy guides the processes and procedures involved in digital forensic investigations [15, 22]. It also provides official recognition of the role of digital forensics in the enterprise [22].

Block A1 is decomposed into several Level B blocks, each of which represents a phase in the digital forensic investigation model of Carrier and Spafford [6]. The phases are incorporated in the framework to highlight the fact that a forensic policy must cover all the investigative phases. Since the focus is on forensic readiness, we only list the incident response phase (Block B2). It is important to note that the decomposition from Level A to Level B is logical, not physical. Thus, each phase of a digital forensic investigation does not require a separate policy; for example, all the phases may be addressed using a single forensic policy (i.e., the overall policy).

The policy in Level B is implemented as procedures or processes in Level C (Figure 1). Because of the focus on digital forensic readiness,

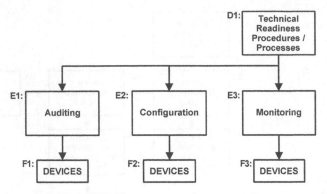

Figure 2. Technical components (Levels D – F).

we only follow the branches leading from Block C1. Block C1 expands to Block D1 (technical readiness procedures and processes) and Block D2 (non-technical readiness procedures and processes).

4.2 Technical Readiness Components

Blocks D1 and D2 represent the technical and non-technical components of digital forensic readiness. According to Rowlingson [19], monitoring and auditing are important components of digital forensic readiness because they help detect and deter incidents. Additionally, procedures and processes must be in place to retrieve and preserve data in an appropriate manner. This is modeled by splitting Block D1 into Blocks E1 through E3 (Figure 2).

Block E2 covers configuration standards, procedures and processes. Blocks E1 and E3 (auditing and monitoring) depend on what is identified under Block E3, and may not be possible unless the hardware and software are configured properly. Consider, for example, two cases: (i) a firewall is not configured to log certain events, and (ii) a firewall and switch are both configured to log events, but are configured to use different time servers. In the first case, events that are not logged by the firewall will not be observed by the monitoring and auditing processes. In the second case, it may be difficult to correlate events from the switch and firewall, which reduces the evidentiary value of the logs that are produced.

Blocks F1 through F3 denote the monitoring, auditing and configuration devices (hardware, software and policy) used in the appropriate business process.

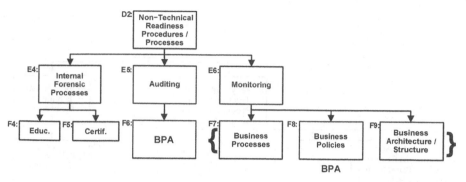

Figure 3. Non-technical components (Levels D – F).

4.3   Non-Technical Readiness Components

The branches from Block D2 in Figure 3 are concerned with the non-technical aspects of digital forensic readiness. Many of the forensic readiness aspects pertinent to privacy are found in this part of the framework. The non-technical components of the framework comprise internal forensic processes, auditing and monitoring (Blocks E4 through E6).

The internal forensic processes in Block E4 are processes that are unique to the forensic team of an enterprise. An example of such a process is the education [14] of forensic team members (Block F4). When implementing a forensic readiness capability for information privacy incidents, it is important to educate forensic investigators (who are primarily trained in security) about information privacy laws. Forensic team members should also have the appropriate certifications (Block F5). These include certifications for conducting digital forensic investigations as well as privacy-related certifications [10].

Blocks E5 and E6 refer to the auditing and monitoring of business processes, policies and architecture. The business processes and policies are those that have relevance to information privacy in the enterprise. Likewise, the business architecture is limited to the structure of the business as it pertains to information privacy. Examples include the creation of a chief privacy officer (CPO) and the creation of a multi-disciplinary team [13] consisting of staff from the office of the CPO, information security, forensics and legal departments. Blocks F7 through F9 correspond to business processes, policies and architecture, respectively. Block F6 expresses the interactions and impact of the business processes, policies and architecture.

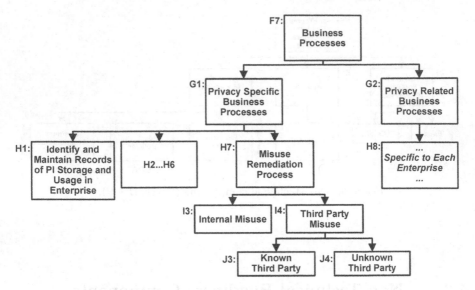

Figure 4. Business processes.

Privacy and Business Processes

Figure 4 shows the decomposition of business processes into privacy-specific and privacy-related business processes from Block F7 to Blocks G1 and G2. Block G2 is an abbreviation of these processes since they are unique to each enterprise and depend largely on the nature of the enterprise. For example, in a delivery company, the process of capturing the details of a delivery to a client is considered to be a privacy-related process because the client's address is private information. Including privacy-related processes in the framework is important because it gives digital forensic investigators immediate information about the business processes likely to be involved in privacy incidents.

Privacy-specific business processes, on the other hand, are processes that deal purely with information privacy. They ensure that the actions required to protect information privacy and enforce the privacy rights of data subjects are in place within the enterprise. The processes are shown as branches of Block G1 in Figure 4. The following processes are omitted to save space: process for communicating the privacy policy (Block H2), process for aligning the privacy policy with business policies (Block H3), process for handling requests to access private information (Block H4), process for correcting private information (Block H5), and process for complaints and complaint escalation (Block H6).

The privacy-specific business processes in the framework are taken from the Generally Accepted Privacy Practices (GAPP) Standard [5].

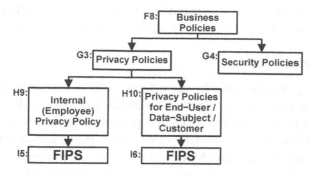

Figure 5. Privacy policies.

Block H7 (misuse remediation process) is used as an example of the many privacy-specific business processes. Misuse remediation describes incidents in which private information is used in a manner that has not been sanctioned by the data subject. Misuse is divided into internal misuse and third party misuse, expressed using Blocks I3 and I4, respectively. The delineation provides for the different digital forensic readiness processes that may be required for each category and sub-category. For example a readiness process for handling privacy incidents with a business partner may include the establishment of a joint forensic team at the outset of the partnership.

Privacy Policies

Figure 5 shows the information privacy policies of an enterprise. Privacy policies in the framework are split into an internal privacy policy for employees of the enterprise (Block H9) and privacy policies for data subjects (Block H10). The internal privacy policy defines guidelines for the acceptable use of private information (belonging to data subjects) by employees. As such, it plays an important role in defining an information privacy incident because an incident usually occurs when the policy has been violated by an employee. It also clarifies the repercussions for employees if they do not adhere to the guidelines.

Privacy policies for data subjects also inform data subjects about the enterprise's practices regarding their private information. Data subjects may then hold an enterprise to the policies and can institute complaints when they believe that the enterprise has not adhered to the policies. The policies are clearly very useful to a forensic investigator tasked with investigating a complaint by a data subject.

In the forensic framework, the internal privacy policy and the privacy policies for data subjects are based on the Fair Information Principles

(FIPS) that underlie most information privacy laws [9]. Other guidelines (e.g., applicable laws) may also be included in Blocks I5 and I6.

5.   Discussion

One of the primary goals in the design of the framework is the inclusion of information privacy protection in the forensic readiness capability of an enterprise. Following the accepted notion that security-related forensic readiness is not possible unless basic information security processes (e.g., logging and incident reporting) are in place [8, 22], we hold that the same is true for a forensic readiness capability for information privacy incidents. An enterprise must implement information privacy practices to maintain a forensic readiness capability for information privacy incidents. The GAPP Standard [5] is used to incorporate specific measures for protecting information privacy within the framework. Enterprises with higher levels of maturity regarding information privacy protection are more likely to have better forensic readiness capabilities for information privacy incidents than those with lower levels of maturity [18].

The framework also incorporates established concepts from security-related forensic readiness [8, 19, 22, 23], namely a policy and a process approach to forensic readiness. Indeed, the primary contributions are the combination of these established concepts with information privacy protection measures and the definition of the relation between the policies, processes and procedures with respect to information privacy incidents. While the principal goal is the inclusion of information privacy protection in the forensic readiness capability of an enterprise, the framework itself is intended to serve as a theoretical guide for developing a forensic readiness capability for information privacy incidents. It is unlikely that the theoretical framework would be implemented "as is" in a real-world enterprise. Policies and processes that exist as separate elements in the framework may be combined if they already exist in an enterprise. Also, an enterprise may omit certain policies and processes. However, this introduces a risk in that certain aspects of information privacy protection may not be covered by the readiness capability. Risk and cost-benefit analyses [19] may be used to determine which, if any, items could be excluded.

A similar exercise to the mapping of technologies to business processes can be conducted with privacy policies and privacy-specific business processes. This could ensure that a digital forensic investigator knows which policies are relevant to incidents that involve specific business processes.

6.   Conclusions

The digital forensic readiness framework for information privacy incidents is motivated by previous work on digital forensic readiness that identifies the need for policies, procedures and processes. It also encompasses information privacy imperatives by drawing on the Fair Information Principles, the GAPP Standard and the information privacy literature. The framework blends concepts from digital forensic readiness and information privacy to provide the essential elements for conducting digital forensic investigations of information privacy incidents. In particular, it provides enterprises with guidance for specifying high-level policies, business processes and organizational functions, and for determining the device-level forensic procedures, standards and processes required to implement a forensic readiness capability for information privacy incidents.

Our future work will refine the framework based on feedback from enterprises with mature forensic readiness capabilities. In addition, an ontology will be used to capture the relationships between framework elements and support automated reasoning.

References

[1] G. Antoniou, L. Sterling, S. Gritzalis and P. Udaya, Privacy and forensics investigation process: The ERPINA protocol, *Computer Standards and Interfaces*, vol. 30(4), pp. 229–236, 2008.

[2] H. Berghel, BRAP forensics, *Communications of the ACM*, vol. 51(6), pp. 15–20, 2008.

[3] H. Burkert, Privacy-enhancing technologies: Typology, critique, vision, in *Technology and Privacy: The New Landscape*, P. Agre and M. Rotenberg (Eds.), MIT Press, Cambridge, Massachusetts, pp. 125–142, 1997.

[4] M. Caloyannides, *Privacy Protection and Computer Forensics*, Artech House, Norwood, Massachusetts, 2004.

[5] Canadian Institute of Chartered Accountants, Generally Accepted Privacy Principles, Toronto, Canada (www.cica.ca/index.cfm/ci_id /258/la_id/1.htm).

[6] B. Carrier and E. Spafford, An event-based digital forensic investigation framework, *Proceedings of the Fourth Digital Forensic Research Workshop*, 2004.

[7] R. Clarke, Introduction to Dataveillance and Information Privacy and Definitions of Terms, Xamax Consultancy, Chapman, Australia (www.rogerclarke.com/DV/Intro.html), 2006.

[8] B. Endicott-Popovsky, D. Frincke and C. Taylor, A theoretical framework for organizational network forensic readiness, *Journal of Computers*, vol. 2(3), pp. 1–11, 2007.

[9] R. Gellman, Does privacy law work? in *Technology and Privacy: The New Landscape*, P. Agre and M. Rotenberg (Eds.), MIT Press, Cambridge, Massachusetts, pp. 193–218, 1997.

[10] International Association of Privacy Professionals, IAPP Privacy Certification, York, Maine (www.privacyassociation.org/index.php?option=com_content&task=view&id=17&Itemid=80).

[11] Y. Jordaan, South African Consumers' Information Privacy Concerns: An Investigation in a Commercial Environment, Ph.D. Thesis, Department of Marketing and Communication Management, University of Pretoria, Pretoria, South Africa, 2003.

[12] S. Lau, Good privacy practices and good corporate governance – Hong Kong experience, *Proceedings of the Twenty-Third International Conference of Data Protection Commissioners*, 2001.

[13] V. Luoma, Computer forensics and electronic discovery: The new management challenge, *Computers and Security*, vol. 25(2), pp. 91–96, 2006.

[14] G. Mohay, Technical challenges and directions for digital forensics, *Proceedings of the First International Workshop on Systematic Approaches to Digital Forensic Engineering*, pp. 155–161, 2005.

[15] M. Noblett, M. Pollitt and L. Presley, Recovering and examining computer forensic evidence, *Forensic Science Communications*, vol. 2(4), 2000.

[16] A. Oliver-Lalana, Consent as a threat: A critical approach to privacy negotiation in e-commerce practices, *Proceedings of the First International Conference on Trust and Privacy in Digital Business*, pp. 110–119, 2004.

[17] Organization for Economic Cooperation and Development, OECD Guidelines on the Protection of Privacy and Transborder Flows of Personal Data, Paris, France (www.oecd.org/document/18/0,3343, en_2649_34255_1815186_1_1_1_1,00.html).

[18] K. Reddy and H. Venter, Privacy Capability Maturity Models within telecommunications organizations, *Proceedings of the Southern African Telecommunication Networks and Applications Conference*, 2007.

[19] R. Rowlingson, A ten step process for forensic readiness, *International Journal of Digital Evidence*, vol. 2(3), 2004.

[20] South African Law Reform Commission, Privacy and Data Protection, Discussion Paper 109, Project 124, Pretoria, South Africa (www.doj.gov.za/salrc/dpapers.htm), 2005.

[21] C. Taylor, B. Endicott-Popovsky and D. Frincke, Specifying digital forensics: A forensics policy approach, *Digital Investigation*, vol. 4(S1), pp. 101–104, 2007.

[22] H. Wolf, The question of organizational forensic policy, *Computer Fraud and Security*, vol. 2004(6), pp. 13–14, 2004.

[23] A. Yasinsac and Y. Manzano, Policies to enhance computer and network forensics, *Proceedings of the Second IEEE Workshop on Information Assurance and Security*, pp. 289–295, 2001.

IV

NETWORK FORENSICS

Chapter 12

TRACKING CONTRABAND FILES TRANSMITTED USING BITTORRENT

Karl Schrader, Barry Mullins, Gilbert Peterson and Robert Mills

Abstract   This paper describes a digital forensic tool that uses an FPGA-based embedded software application to identify and track contraband digital files shared using the BitTorrent protocol. The system inspects each packet on a network for a BitTorrent Handshake message, extracts the "info hash" of the file being shared, compares the hash against a list of known contraband files and, in the event of a match, adds the message to a log file for forensic analysis. Experiments demonstrate that the system is able to successfully capture and process BitTorrent Handshake messages with a probability of at least 99.0% under a network traffic load of 89.6 Mbps on a 100 Mbps network.

Keywords: Peer-to-peer file sharing, BitTorrent, forensic tool, packet analysis

1.   Introduction

The use of the Internet for peer-to-peer (P2P) file sharing has steadily increased since Napster was introduced in 1999. A 2000 University of Wisconsin study [9] found that Napster traffic had supplanted HTTP traffic as the dominant protocol used in the university's network. In 2002, researchers at the University of Washington [10] determined that P2P traffic accounted for 43% of all university traffic, with only 14% of traffic devoted to HTTP. Another study [2] found that 50-65% of downloads and 75-90% of uploads were P2P related. A 2005 survey [2] estimated that P2P networks contained more than 2.8 billion downloadable files. According to a Cachelogic report [11], approximately 61% of all Internet traffic is P2P related, compared with only 32% for HTTP.

The long-term goal of our research is to develop a system that identifies and tracks any type of digital file that is transmitted on a network using P2P protocols. The final system will consist of a suite of tools

G. Peterson and S. Shenoi (Eds.): Advances in Digital Forensics V, IFIP AICT 306, pp. 159–173, 2009.

to detect P2P transmissions on a target network, classify them according to the P2P protocol used, compare the digital file being transmitted against a contraband list, and identify the sender and recipient by their IP addresses. This system, implemented as a digital forensic tool, will enable a user to monitor network traffic in real-time for files shared via P2P protocols that meet the user's definition of contraband. Therefore, the system should be of great interest to systems administrators as well as law enforcement personnel. Law enforcement agents could use the system to identify child pornography being transmitted across a network, and track the sender and receiver to their sources.

The rest of this paper is organized as follows. Section 2 discusses methods for tracking illegal file sharing and describes the BitTorrent P2P protocol. Section 3 describes the process used to build our Field Programmable Gate Array (FPGA)-based forensic tool that detects BitTorrent packets and matches the files being shared to a contraband list. Section 4 discusses the experiments used to evaluate the ability of the tool to capture and analyze packets at near line speed. Section 5 presents the experimental results and analysis, and Section 6 presents our conclusions and directions for future work.

2.   Related Work

This section describes methods for identifying illegal file sharers and the popular BitTorrent protocol, which is the focus of our work.

2.1   Identifying Illegal File Sharers

Given the rapid increase in P2P file sharing, law enforcement agencies and copyright holders are struggling to identify illegal file sharers. Several methods are available for identifying and tracking illegal file downloaders. One approach is to use honeypots. A newer method, which is used to identify illegal downloads on BitTorrent, involves the exhaustive search of tracker servers.

Honeypots In the context of this discussion, a honeypot is a trap designed to detect and track illegal file sharing activities. The most basic form of a honeypot involves setting up a computer with a collection of illegal files on the Internet. When another computer attempts to download the illegal files, the downloader's IP address and port number, the date and time of the download, and the downloaded packets are recorded by the honeypot.

Badonnel, *et al.* [1] have developed a management platform for tracking illegal file sharers in P2P networks using honeypots. However, there are some shortcomings to using honeypots for identifying and tracking illegal file sharers. In order to be effective, the file sharer must be able to find and access the honeypot. To prevent this, programs such as Peer Guardian contain blacklists of IP addresses known to contain honeypots and prevent the user's P2P software from downloading files from these blacklisted sites [5]. Another shortcoming is that the use of a honeypot represents an active method of detection – file sharers must download from the honeypot in order to be identified by law enforcement agencies. In the case of highly illegal files (e.g., child pornography), private invite-only websites and/or hard-to-locate websites help keep away members of the general public and law enforcement agents [7].

BitTorrent Monitoring System   The BitTorrent Monitoring System (BTM) [2] can also be used to detect and track illegal file downloaders. BTM automatically searches for BitTorrent-based downloadable files, analyzes the files to determine if they are illegal, attempts to download the suspected illegal files, and records tracking information about the computer that provided the files for download.

BTM has the potential to become a powerful tool for combating illegal file sharing. However, the system has some drawbacks. First, due to the massive number of files that are available on most BitTorrent websites, BTM currently has a very slow processing time. As the number of sublevels covered by the search algorithm increases, the number of total `.torrent` files to be analyzed increases exponentially. Because it cannot run in real time, BTM is unable to cope with the constantly-changing peer lists produced by the tracker sites being monitored.

2.2   BitTorrent Protocol

This paper focuses on the BitTorrent protocol [4]. BitTorrent differs from other distributed P2P protocols in that it allows downloaders to obtain pieces of files from tens or hundreds of other users simultaneously. To further speed up downloads, any user who downloads pieces of files also uploads those pieces he already possesses. The protocol achieves very high download rates by aggregating the slower upload speeds of hundreds of peers [3].

The key BitTorrent component used in this research is the "info hash" of the file dictionary, which is found in the `.torrent` file that contains metadata about the data to be shared. To create the info hash, the SHA-1 algorithm [8] is applied to the information dictionary contained in the `.torrent` file. The resulting message digest is labeled as the "file info hash," which uniquely identifies the file offered for download regardless of the file description in the `.torrent` file. The client provides the file

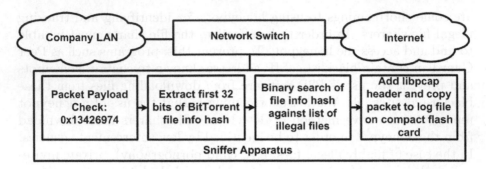

Figure 1.   Packet data flow through the forensic tool.

info hash as the file identifier in the request for a peer list and also when establishing connections using the Handshake message. By comparing this hash value against a list of hashes compiled from the .torrent files associated with the data of interest, it is possible to determine if the client is attempting to share a file on the contraband list.

3.   Forensic Tool

The goal of this research is to develop an FPGA-based embedded software system that allows for the capture and evaluation of Ethernet packets transmitted on a LAN and between the LAN and the Internet. The FPGA implementation enables the software application to directly access the Ethernet controller buffers, bypassing the rest of the network stack and enhancing system simplicity and speed.

Figure 1 shows the packet data flow through the forensic tool. When a packet enters the system, the first 32 bits of the payload are extracted and compared with the first 32 bits of a valid BitTorrent Handshake message, which is 0x13426974. The frame is discarded if the first word of the payload of the frame does not match this string. If the word does match, the first 32 bits of the info hash of the Handshake packet's file are extracted from another location in the frame, and compared against a list of hashes belonging to files of interest. If the file info hash is not in the list, the frame is dropped. If the file info hash is in the list, the frame is saved in a Wireshark-readable log file and placed on a compact flash card. The frames recorded in the log file are subsequently analyzed to extract IP address information for tracking and forensic analysis.

3.1   Initial System Configuration

The current prototype is implemented as an embedded software application using the Power PC core of a Virtex II Pro FPGA development board. Xilinx-supplied drivers and built-in functions are used where possible, with custom software used to accomplish certain functions: loading the data file containing the file info hashes of the contraband data, performing packet payload inspections, copying BitTorrent Handshake frames to on-chip RAM, comparing hash values, and writing frame data to the compact flash card.

The salient features of the prototype are:

- All the modules are implemented in software. However, the hardware is modified to enable the Ethernet controller to operate in the promiscuous mode.

- Packets of interest are copied three times. The first is from the Ethernet controller to RAM upon detection of the 32-bit BitTorrent signature in the packet payload. If the file info hash is found in the list, the frame is copied from RAM to a character array, and then from the array to the log file on the compact flash card.

- Frames are copied to the compact flash card as they are processed. The system waits until the current packet has been processed completely and sent to the compact flash card before it processes another packet.

3.2   System Optimization

The following optimizations were investigated to improve the performance of the prototype (i.e., increase packet processing speed):

- Removing all user notifications of packets found using the serial port and HyperTerminal ("User Alerts" configuration): Because the serial port runs at a much lower speed than the CPU and processing bus, it is hypothesized that sending data over the RS-232 connection dramatically increases the overall processing time.

- Storing all captured frames of interest within a RAM block instead of writing them individually to the compact flash card ("Packet Write" configuration): By storing the data within RAM, write functions to the compact flash card are only performed before packet sniffing begins and after packet sniffing terminates. It is hypothesized that writing to the compact flash card is a high-latency process and eliminating it saves a significant amount of processing time.

- Adding a second receive buffer to the Ethernet controller ("Dual Buffer" configuration): This enables one frame to be processed while the next frame is received: The goals are to give the com-

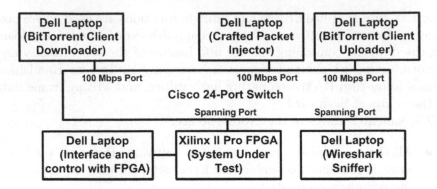

Figure 2.   Experimental setup.

parison and copying routines additional time to execute, and to limit the number of frames dropped due to a full receive buffer.

- Enabling the instruction and data caches of the Power PC processor ("Cache" configuration): It is hypothesized that allowing the FPGA to cache processor instructions, heap data and stack data instead of performing multiple reads and writes to block RAM results in significant processing time savings.

- Integrating the four optimization techniques in a single system ("Combined" configuration): The goal is to leverage each optimization individually and to gain synergistic time savings by combining all four optimizations.

4.   Testing Methodology

Figure 2 shows the experimental setup used to test the various configurations and validate the system design. The experimental setup incorporates two Dell Inspiron Windows XP laptops loaded with uTorrent, a popular BitTorrent client, and a Dell Inspiron Linux laptop configured with the hping utility to inject crafted BitTorrent Handshake packets. The three laptops are connected to a Cisco Catalyst 2900XL 100 Mbps switch. Our Virtex II Pro FPGA system is connected to a spanning port on the switch. One Dell Inspiron Windows XP laptop loaded with Wireshark is placed on a second spanning port as a control packet analyzer. The other Dell Windows XP laptop is used to configure and load the Virtex II Pro via a USB port and to receive alerts through a HyperTerminal connected via serial port. A data file containing 1,000 file info hashes is used as the list of interest in our experiments.

Two experiments were conducted. The first experiment recorded the numbers of cycles required to process three types of packets. The second experiment tested the ability of the system to detect and process Handshake packets with the network running at near maximum capacity.

4.1   Packet Processing Time

The first test involved sending a series of packets from the Crafted Packet Injector to the BitTorrent Client Downloader. A series of 50 frames were sent for each of three types of packets and the CPU cycles needed to process the packets were recorded. The three types of packets tested were: (i) packets that did not correspond to BitTorrent Handshake messages, (ii) Handshake message packets whose file info hash values were not in the list of interest, and (iii) Handshake messages whose file info hash values were in the list of interest. The three test packet series were created by extracting the payloads from a series of actual BitTorrent file transfers, copying the payload contents into a binary file with a hex editor, and using the hping utility to inject exactly 50 of each type of copied packet into the network.

The following configurations were tested in order to assess the improvement provided by each optimization technique: Control (software-only implementation with no user alerts), Control with User Alerts, Packet Write, Dual Receive Buffer, Cache and Combined.

4.2   Probability of Intercept Under Load

The second test involved sending a series of BitTorrent Handshake messages from the Crafted Packet Injector to the BitTorrent Client Downloader with the network under a heavy load. To create the load, a 1.5 GB video file was transferred from the BitTorrent Client Uploader laptop to the BitTorrent Client Downloader laptop using the Windows NETBIOS file transfer protocol. While the download was in progress, a series of 300 BitTorrent Handshake messages (whose file info hash values were in the list) were sent 0.2 seconds apart to the BitTorrent Client Downloader laptop using the hping utility. Because the packets were injected 0.2 seconds apart, the results of each trial (either the packet was captured or not captured) can be assumed to be independent of each other. At the end of the test, the number of packets successfully written to the log file was recorded for each configuration.

To measure the minimum overall network load, the Wireshark utility was used to analyze all the traffic sent during the test and to compute the average network load. The configurations used in the first test were used to assess the improvement provided by each optimization technique. To permit better comparisons, this test also used the Wireshark packet analyzer tool.

Table 1. Packet processing times for non-BitTorrent packets.

Configuration	Mean	Percent Change	Standard Deviation	Confidence Interval (95%)
Control	1,206	0.00	0.00	(1,206, 1,206)
User Alerts	1,152	4.48	0.00	(1,152, 1,152)
Dual Buffer	1,344	(11.44)	109.10	(1,313, 1,375)
Packet Write	1,146	4.98	0.00	(1,146 1,146)
Cache	276	77.11	0.00	(276, 276)
Combined	303.5	74.83	25.76	(296.18, 310.82)

5.   Results and Analysis

This section presents the results obtained with respect to packet processing times and packet interception probabilities under network load, along with the accompanying analysis.

5.1   Packet Processing Times

Table 1 presents the results of one-variable t-tests performed for the six configurations using the non-P2P packet type. For each configuration, the table lists the mean number of CPU cycles required to process non-P2P packets, the percent change in processing time from the Control configuration, the standard deviation, and the 95% confidence interval for the mean. The number of cycles required ranges from 276 cycles to 1,344 cycles, which equates to a range of 0.92 to 4.48 microseconds per packet. As shown in the table, the addition of a second receive buffer requires additional processing time; all the other configurations require fewer cycles. Note that a significant number of cycles are saved by enabling the instruction and data caches.

Table 2 presents the results of one-variable t-tests performed for the six configurations using BitTorrent Handshake packets whose file info hash values were not in the list of interest. For each configuration, the table lists the mean number of CPU cycles required to process the Bit-Torrent packets, the percent change in processing time from the Control configuration, the standard deviation, and the 95% confidence interval for the mean. The number of cycles required ranges from 1,145 cycles to 7,770 cycles, which equates to a range of 3.82 to 25.9 microseconds per packet. The second receive buffer and the alternate packet writing method require additional processing time; all the other configurations

Table 2. Packet processing times for BitTorrent packets not in the list.

Configuration	Mean	Percent Change	Standard Deviation	Confidence Interval (95%)
Control	7,296	0.00	0.00	(7,296, 7,296)
User Alerts	1,044,756	(14,219.60)	730	(1,044,549, 1,0449,63)
Dual Buffer	7,770	(6.50)	0.00	(7,770, 7,770)
Packet Write	7,593	(4.07)	0.00	(7,593, 7,593)
Cache	1,145	84.31	0.00	(1,145, 1,145)
Combined	1,205	83.48	0.00	(1,205, 1,205)

require fewer cycles. Once again, a significant number of cycles are saved by enabling the instruction and data caches.

Table 3. Packet processing times for BitTorrent packets in the list.

Configuration	Mean	Percent Change	Standard Deviation	Confidence Interval (95%)
Control	116,207	0.00	22,418	(109,836, 122,578)
User Alerts	1,702,125	(1,364.74)	22,880	(1,695,623, 1,708,628)
Dual Buffer	118,986	(2.39)	22,391	(112,623, 125,350)
Packet Write	53,034	54.36	1,146	(52,708, 53,360)
Cache	14,679	87.37	2,064	(14,093, 15,266)
Combined	9,125	92.15	108.8	(9,093.7, 9,155.5)

Table 3 presents the results of one-variable t-tests performed for the six configurations using BitTorrent Handshake packets whose file info hash values were in the list of interest. For each configuration, the table lists the mean number of CPU cycles required to process the BitTorrent packets, the percent change in processing time from the Control configuration, the standard deviation, and the 95% confidence interval for the mean. The number of cycles required ranges from 9,125 cycles to 118,986 cycles, which equates to a range of 30.42 to 396.62 microseconds per packet. The second receive buffer requires additional processing time; all the other configurations require fewer cycles. Note that the Packet Write configuration requires fewer CPU cycles than the other configurations; this is because it is the only test where packets were written to the log file.

The following observations can be made based on the data:

- Adding user alerts significantly increases the processing time for BitTorrent packets. This is because user alerts are transmitted via a serial port at 115,200 baud, which is much slower than the 300 MHz processor speed and 100 MHz bus speed used by the FPGA.

- Adding a second receive buffer increases the number of CPU cycles required to process a packet regardless of the type of packet. The additional processing cycles are required to check both the receive buffers in order to determine which buffer contains the next packet to be processed. However, as discussed in Section 5.2, the increase in CPU cycles is more than offset by the benefits obtained by introducing the second receive buffer.

- As expected, modifying the packet writing routine only decreases the number of CPU cycles required to process packets when packets are actually written to the log file. No significant processing time is gained or lost with this optimization technique when packets are not written.

- Enabling the instruction and data caches produces a significant reduction in the number of CPU cycles required to process packets regardless of packet type.

5.2   Packet Intercept Probabilities Under Load

Table 4 presents the results of the packet intercept test under a heavy network load. In particular, the table shows the number of packets captured out of the 300 sent packets for each configuration. The probability of intercept and the corresponding 95% confidence interval are also shown for each configuration. In all the tests, the total load on the network as measured by the Wireshark packet analyzer was between 89.6 Mbps and 89.7 Mbps, which equates to a 90% load (approx.) on the 100 Mbps network. However, this measurement is not absolute because Wireshark can drop packets under a heavy load. Since it is not known how many packets were actually dropped by Wireshark, we consider 89.6% to be the minimum load on the test network.

The results in Table 4 demonstrate that while the User Alerts and Packet Write configurations capture more packets of interest than the Control configuration (166 and 174 versus 159), the overlapping confidence intervals suggest that the differences are not statistically significant. Also, the Cache and Dual Buffer configurations perform signifi-

Table 4. Packet intercept probability under high network load.

Configuration	Packets Captured	Packets Sent	Probability of Capture	Confidence Interval (95%)
Control	159	300	0.5300	(0.4718, 0.5876)
User Alerts	166	300	0.5533	(0.4951, 0.6105)
Packet Write	174	300	0.5800	(0.5219, 0.6365)
Cache	289	300	0.9633	(0.9353, 0.9816)
Dual Buffer	292	300	0.9733	(0.9481, 0.9884)
Combined	300	300	1.0000	(0.9901, 1.0000)
Wireshark	298	300	0.9933	(0.9761, 0.9992)

cantly better than the Control configuration. Moreover, the Combined configuration performs the best – a 100% capture rate for packets of interest. Using a 95% confidence interval, this equates to a minimum capture rate of 99.0%; this rate is comparable to the performance of Wireshark, which yielded a minimum capture rate of 97.6%.

Table 5. Hypothesis testing (Control): Packet intercept under high network load.

Alternative Hypothesis with 95% Confidence Interval	Estimate for Difference	Z Value of Diff. Test	P Value of Diff. Test
p(User Alerts) > p(Control)	0.0233	0.57	0.283
p(Packet Write) > p(Control)	0.0500	1.23	0.109
p(Cache) > p(Control)	0.4333	14.07	0.000
p(Dual Buffer) > p(Control)	0.4433	14.64	0.000
p(Combined) > p(Control)	0.4700	16.31	0.000

To further assess the statistical significance of the results, we performed a hypothesis test between each configuration and the Control configuration. As shown in Table 5, the p-value for the one-sided test involving the User Alerts and Control configurations is too high (0.283) to state with confidence that the increase in the probability of intercept is statistically significant. In the one-sided test involving the Packet Write and Control configurations, the p-value is again too high (0.109) to reject the hypothesis outright; however, it can be inferred that there is some improvement in the probability of intercept. Finally, p-values of

Table 6. Hypothesis testing (Combined): Packet intercept under high network loads.

Alternative Hypothesis with 95% Confidence Interval	Estimate for Difference	Z Value of Diff. Test	P Value of Diff. Test
p(Combined) > p(User Alerts)	0.4467	15.56	0.000
p(Combined) > p(Packet Write)	0.4200	14.74	0.000
p(Combined) > p(Cache)	0.0367	3.38	0.000
p(Combined) > p(Dual Buffer)	0.0267	2.87	0.002
p(Combined) > p(Wireshark)	0.0067	1.42	0.078

0.000 are obtained for the one-sided tests for the Cache, Dual Buffer and Combined configurations. Thus, a strong statistical certainty exists that each of these configurations is better than the Control configuration.

To determine the overall performance of the Combined configuration, hypothesis tests were performed for the Combined configuration versus the individual optimizations and Wireshark. As shown in Table 6, the p-values for the one-sided tests involving the User Alerts, Packet Write, Cache and Dual Buffer configurations range between 0.000 and 0.002, indicating a strong statistical certainty that the Combined configuration is better than each individual optimization. For the performance of Wireshark versus the Combined configuration, Table 6 shows that the p-value for the one-sided test is 0.078, which is too high to reject the hypothesis; but it still indicates that the two have comparable performance.

5.3 Analysis of Results

The most significant reduction in the number of CPU cycles needed to process packets of interest occurs when the data and instruction caches are enabled for the Power PC processor. By allowing the FPGA to cache processor instructions as well as heap and stack data, the packet processing time is reduced by 77% to 84% depending on packet type. In addition, by delaying the compact flash write operations until after sniffing has terminated, the packet processing time is reduced by 54% for packets written to the log file. When all four optimizations are combined, a 74% to 92% improvement is obtained in the packet processing time over the Control configuration (depending on packet type).

The significant packet loss rate for the single receive buffer configurations in the packet capture tests is likely due to the inability of an Ethernet frame to be processed and cleared from the buffer before the next frame arrives. At 100 Mbps, the mandatory interframe gap required by the Ethernet protocol produces a 0.96 microsecond delay between frames. Because multiple instructions are required to transfer data from the Ethernet buffer, read the payload contents and analyze the data, the system – which can perform at most 300 instructions per microsecond – cannot keep up with the data flow. This results in significant packet loss as the system approaches 100% utilization. However, it is important to note that this observation does not hold for the Cache configuration: enabling the caches provides a capture rate of 96%, even in the case of a single buffer. This is likely due to the fact that the extremely small processing times provided by the cache enable packets to be processed in the short interframe time gap.

Adding a second receive buffer to the Ethernet controller dramatically increases the probability of packet intercept under load – a 97% capture rate even with no other optimizations. The use of two receive buffers enables a packet to be processed from one buffer while the next packet is being received in the other buffer. Specifically, the additional buffer provides a minimum of 576 additional bit times ((7-byte preamble + 1-byte delimiter + 64-byte minimum frame size) × 8 bits/byte) [6] for processing each frame over the single buffer option. Although this improvement comes at the cost of additional processing cycles, the expanded processing window provided by the second buffer more than offsets the cost incurred by individual packet processing. When combined with caching and an improved packet writing scheme, the infrequency of packets of interest and the small likelihood of traffic saturation on the network link, the final design allows the system to successfully capture and process all the packets of interest on the wire.

6. Conclusions

This paper has described the design of a specialized forensic tool that uses a Virtex II Pro FPGA to detect BitTorrent Handshake packets, compare the packets' file info hash values against a list of hashes preloaded into memory, and in the event of matches, and save the packets in a log file for further analysis. Several optimization techniques for reducing the CPU time required to process packets are investigated, along with their ability to improve packet capture performance. The results demonstrate that the fully optimized forensic tool can intercept, process and store packets of interest with a minimum of 99.0% probability of success even under heavy network load.

The next step in our research is to extend the system to include other P2P protocols while maintaining its overall speed and accuracy. Specifically, we are focusing on the Session Initiation Protocol (SIP), which is widely used in Voice-over-IP applications. In addition, we plan to investigate system performance at higher network speeds using a gigabit network and Xilinx Virtex-5, a more powerful FPGA board. Our future research will also focus on message stream encryption and protocol encryption capabilities of BitTorrent clients.

References

[1] R. Badonnel, R. State, I. Chrisment and O. Festor, A management platform for tracking cyber predators in peer-to-peer networks, *Proceedings of the Second International Conference on Internet Monitoring and Protection*, p.11, 2007.

[2] K. Chow, K. Cheng, L. Man, P. Lai, L. Hui, C. Chong, K. Pun, W. Tsang, H. Chan and S. Yiu, BTM – An automated rule-based BT monitoring system for piracy detection, *Proceedings of the Second International Conference on Internet Monitoring and Protection*, p. 2, 2007.

[3] B. Cohen, Incentives build robustness in BitTorrent (www.bittor rent.org/bittorrentecon.pdf), 2003.

[4] B. Cohen, BEP3: The BitTorrent protocol specification (www.bittor rent.org/beps/bep_0003.html), 2008.

[5] P. Gil, "Peer Guardian" Firewall: Keep your P2P private (netfor beginners.about.com/od/peersharing/a/peerguardian.htm), 2009.

[6] Institute of Electrical and Electronics Engineers, IEEE Standard 802.3-2005: Local and Metropolitan Area Networks – Specific Requirements Part 3: Carrier Sense Multiple Access with Collision Detection (CSMA/CD) Access Method and Physical Layer Specifications, Piscataway, New Jersey (standards.ieee.org/getieee802 /802.3.html), 2005.

[7] R. MacManus, The underground world of private P2P networks (www.readwriteweb.com/archives/private_p2p.php), 2006.

[8] National Institute of Standards and Technology, Secure Hash Standard (FIPS 180-1), Federal Information Processing Standard Publication 180-1, Gaithersburg, Maryland (www.itl.nist.gov/fipspubs /fip180-1.htm), 1995.

[9] D. Plonka, UW-Madison Napster traffic measurement, University of Wisconsin, Madison, Wisconsin (net.doit.wisc.edu/data/Napster), 2000.

[10] S. Saroiu, K. Gummadi, R. Dunn, S. Gribble and H. Levy, An analysis of Internet content delivery systems, *Proceedings of the Fifth Symposium on Operating Systems Design and Implementation*, pp. 315–327, 2002.

[11] TorrentFreak, The "one-third of all Internet traffic" myth (torrentfreak.com/bittorrent-the-one-third-of-all-internet-traffic-myth), 2006.

Chapter 13

A MODEL FOR FOXY PEER-TO-PEER NETWORK INVESTIGATIONS

Ricci Ieong, Pierre Lai, Kam-Pui Chow, Frank Law, Michael Kwan and Kenneth Tse

Abstract   In recent years, peer-to-peer (P2P) applications have become the dominant form of Internet traffic. Foxy, a Chinese community focused file-sharing tool, is increasingly being used to disseminate private data and sensitive documents in Hong Kong. Unfortunately, its scattered design and a highly distributed network make it difficult to locate a file originator. This paper proposes an investigative model for analyzing Foxy communications and identifying the first uploaders of files. The model is built on the results of several experiments, which reveal behavior patterns of the Foxy protocol that can be used to expose traces of file originators.

Keywords: Peer-to-peer network forensics, Foxy network, Gnutella 2 protocol

1.   Introduction

Recent surveys report that P2P traffic is responsible for 41% to 90% of all Internet traffic [2, 4]. In 2007, two popular P2P file-sharing applications, BitTorrent and eDonkey, contributed 50% to 70% and 5% to 50% of all P2P traffic, respectively [2]. The Foxy P2P file-sharing protocol is gaining popularity in traditional Chinese character markets such as Hong Kong and Taiwan – approximately 500,000 users are active on the Foxy network at any given time [14]. A Foxy client, which is available free-of-charge, provides a user-friendly traditional Chinese interface. It enables users to connect to the Foxy network without any special configuration and to download free music, movies and software with just a few keystrokes and mouse clicks.

Foxy drew worldwide attention when hundreds of photographs of a local pop icon participating in sex acts with female celebrities were dis-

G. Peterson and S. Shenoi (Eds.): Advances in Digital Forensics V, IFIP AICT 306, pp. 175–186, 2009.

seminated on the Internet [5]. Other cases involving Foxy include the unintentional sharing of sensitive files in Taiwan and Hong Kong [12]. Locating a file originator is very difficult due to Foxy's distributed network. While Hong Kong Customs and Excise are able to identify and prosecute BitTorrent seeders [3], the techniques used for BitTorrent are not applicable to Foxy.

Chow, *et al.* [7] have published guidelines on the secure use of Foxy clients. However, to our knowledge, no publication describes an investigative process that is applicable to the Foxy network. Without a mechanism to identify file originators, it is impossible to collect digital evidence for prosecuting illegal publishers of digital materials as in the pop icon scandal. This paper addresses the issue by proposing an investigative model for analyzing Foxy communications and identifying the first uploaders of files.

2.   P2P Forensic Tools

Several digital forensic tools have been developed to identify traces of P2P client execution on computers [1, 16]. However, locating the original uploaders of files remains one of the most challenging problems in P2P network forensics.

The BTM tool [6] was designed to monitor the BitTorrent network and identify initial seeders. BTM monitors public web forums where BitTorrent users communicate and announce their torrent files (the seed files for file download in the network). It mimics a BitTorrent client and collects communication information from trackers and peers. Initial seeders are identified by analyzing the collected data.

However, the BTM approach cannot be applied to the Foxy network because Foxy clients use the Gnutella 2 protocol [9]. Moreover, the mechanisms used to broadcast shared files are different. In BitTorrent, a torrent file is published on a web forum for others to download; no auxiliary seed file is needed in the case of the Foxy network.

Nasraoui, *et al.* [13] have proposed a node-based probing and monitoring mechanism for Freenet and Gnunet. Their approach bears some similarity to our method. However, they only provide a high level framework and do not address the issue of locating initial uploaders.

3.   Foxy Network Overview

Foxy is a hybrid P2P model based on the Gnutella 2 protocol (G2). In the Foxy network, a peer is either an "ultrapeer" or a "leaf node." Ultrapeers are active nodes that coordinate surrounding nodes, filtering and directing query traffic within the Foxy network. Leaf nodes are the most common nodes; they issue search queries, upload and download files. Ultrapeers are used to relay communications from leaf nodes without flooding the Foxy network.

The search mechanism in the G2 network is initiated when a user enters keywords and clicks the "Search" button. The corresponding search query, referred to as a *Q2* packet, is then submitted to the connected ultrapeers. Each ultrapeer verifies and validates the search, and selectively redirects the query to other leaf nodes or ultrapeers. Nodes that are unlikely to have files matching the keywords may not receive the query.

Ultrapeers use a "query hit table" (QHT) to keep track of the files available on its leaf nodes. For each shared file, the file name is hashed using the QHT algorithm [9]. The algorithm hashes both the file name and the keywords that are maintained separately in the QHT. Whenever a QHT entry matches a query either partially or completely, the ultrapeer considers that the corresponding node has the potential to answer the query and forwards the query to that node.

If the receiving node possesses a file with a name matching the received query, a "query hit packet" (called a *QH2* packet) is transmitted back to the requester. The *QH2* packet contains the IP address of the file source as well as the descriptive name of the file. If several nodes respond, the ultrapeer consolidates the results and sends the requestor a newly constructed *QH2* packet. At this point, the requester can initiate a file download based on the search results.

4. Foxy Protocol Analysis

In order to determine and analyze the behavior of the Foxy network, traffic generated and received by several Foxy clients was captured using Wireshark [15], a popular network packet capturing tool. This section describes the experimental results based on the analysis of more than 80 sets of Foxy communication network traffic records involving approximately 3 million packets.

4.1 Data Collection

Table 1 lists the five data collections (A through E) used to analyze the Foxy protocol. The collections, which are of varying lengths, were executed over a five-month period. Data collections A and B focus on the search results of popular keywords. Data collection C compares and analyzes the search results of a query between two Foxy clients. Data collections D and E investigate how search queries and results propagate across multiple clients.

Table 1. Summary of experiments and tested queries.

DC	Date	Purpose	Query	Sets
A	05/14/08 to 06/10/08	Identify the newly-announced keyword (**Pol Record**) and analyze the relevant search results.	Pol Record	10
B	09/11/08 to 09/29/08	Analyze the *Q2* and *QH2* packets of a newly-announced keyword and compare the results with those from data collection A. Observe the connectivity of Foxy peers.	yoshinoya (in Chinese)	40
C	09/29/08	Monitor and analyze *Q2* and *QH2* packets.	mov00423	7
D	09/29/08 to 10/08/08	Investigate the propagation of *Q2* and *QH2* packets among multiple clients.	bhb_od2, mov00423, foxy testing2209, foxy testing1404, 04c3a2d1, 4b9277c6, ad28ae6e, yoshinoya (in Chinese)	24
E	10/09/08	Analyze the *Q2* and *QH2* packets of a self-published file.**Foxy_testing2209.mp3**	foxy testing 2209	6

4.2   Experimental Results

This section describes the experimental results obtained during the analysis of the Foxy network.

Ultrapeer List Changes In the case of data collection B, the Foxy client was connected and disconnected from the Foxy network several times. By analyzing the network packets, it was observed that the client connected to a random list of available ultrapeers provided by the Foxy Gnutella web cache. Since the web cache arbitrarily selects ten to fifteen ultrapeers from the ultrapeer pool, leaf nodes connect to different sets of ultrapeers regardless of their physical or network locations. Also, different lists of ultrapeers are returned to a leaf node whenever it is reconnected to the Foxy network regardless of the length of the disconnection.

Ultrapeer and Leaf Node Connectivity   According to the specifications, ultrapeers have a maximum connectivity of 200 nodes. In the case of data collection B, no ultrapeers were found to exceed this value. On the other hand, leaf nodes were found to be connected to no more than 32 ultrapeers at a time. Because the pool of ultrapeers changes over time, the ultrapeers to which a leaf node is connected also changes over time.

At any time, there are about 500,000 nodes connected to the Foxy network. If each ultrapeer and leaf node is connected to the maximum of 200 leaf nodes and 32 ultrapeers, respectively, then there are at least 80,000 ultrapeers in the network.

Query Results after Reconnections   In the case of data collection D, five files (F1, F2, F3, F4 and F5) with uncommon names were shared using a Foxy client. Another client (at a remote site) was launched to search for these files. In the first connected session, three files (F1, F3 and F4) were returned in the search results. However, after restarting the client, only F2 was located. This shows that different results are received when identical queries are submitted after a reconnection. The likely cause is that, when new files appear, propagation through the ultrapeers takes a period of time and the propagation is potentially limited to nearby peers.

Searching for Files with Uncommon Names   In the case of data collection C, several queries were executed with the keyword MOV00423. This keyword was part of the name of a video file in a rape case [11]. Before the incident, no requester would expect to receive results for this query. However, after the newspapers reported the case, many results were returned for this keyword query.

These results and those for data collection D suggest that files with uncommon file names are difficult to find at the beginning of the file-sharing period. Only after the keyword is queried a number of times are more results returned.

Query Filtering at Ultrapeers   In the case of data collection E, a leaf node X with the file Foxy-testing220.mp3 was connected to the Foxy network. Then, another leaf node Y was connected to the Foxy network; this leaf node submitted a query for the file. After two minutes, the client was disconnected and reconnected to the Foxy network.

Upon analyzing the $Q2$ packets arriving at X, we discovered that the exact matching query Foxy-testing220 was sent through the ultrapeers. However, alternatives to the query such as Foxy-testing22 and

Table 2. Percentage of Q2 packets querying H_S in all Q2 packets

Date	Duration (min)	Q2 Packets (N_1)	Q2 Packets Querying H_S (N_2)	N_2/N_1 (%)
11/09/08	114	833	767	92.1
12/09/08	110	1,097	1,043	95.1
13/09/08	27	219	201	91.8
14/09/08	58	419	357	85.2
18/09/08	22	344	285	82.8
08/10/08	27	560	386	68.9

`Foxy-testing2209` did not arrive at X. This indicates that the filtering performed at the ultrapeers does not implement a sub-string match.

It was observed that query filtering does not employ exact matching. Of the 559 Q2 packets that arrived at X, only 26 were initiated by Y requesting the target file. In fact, 95% of the queries were not for the file being shared, indicating that filtering does not use an exact-matching method.

Approximately 70% of the irrelevant queries contained identical hash values or very similar file names. In other words, the majority of the queries were dominated by a small set of extremely popular keywords that were similar to each other.

Hash Querying after an Incident   Table 2 compares the Q2 packets captured from September 11, 2008 to September 14, 2008; on September 18, 2008; and on October 8, 2008 in data collections B, D and E, respectively. The majority of the queries pertained to a specific hash value, H_S, which was associated with the sex scandal mentioned above.

The large number of Q2 packets querying H_S suggests that popular queries may occupy the query hash table after the announcement of keywords. Fewer queries for H_S appear after the tide passes and/or more peers have downloaded the file,

5.   Hypothesized Foxy Network Behavior

Apart from the findings discussed above, some important behavior patterns were not observed. These behavior patterns, which pertain to the publication of a new file, searching for file keywords and massive P2P downloading of a file, are critical to the success of any P2P file monitoring and investigation technique. Therefore, hypotheses have to be derived based on known features of the protocol.

In any file-sharing environment, there are five stages before a down-loader obtains a file: preparation, initiation, publication, waiting and downloading [10]. The publication, waiting and downloading stages are the only stages where an investigator can collect digital evidence remotely on the Internet, i.e., before examining the suspect machine.

5.1   Publication Stage

After the preparation and initiation stages, the existence of a new file must be announced to the world before it can be disseminated widely. Because the Foxy network does not provide a mechanism for publishing newly-shared files, a file uploader must make the announcement independent of the Foxy network. Therefore, no phenomena can be observed in the Foxy network during the initial publication stage.

5.2   Waiting Stage

The waiting stage is critical in the Foxy network protocol. Shortly after the publication of a new file, Foxy users search for the file by submitting keywords. If the keywords arouse sufficient publicity, a sudden increase in the number of queries asking for the same keyword are observed in the Foxy network. This is partially reflected in the comparison of the queries in data collections A and B.

5.3   Downloading Stage

File sharing completes with the transfer of the entire file from one peer to another. In the ideal case, the file can be distributed to nk peers in k rounds, where n is the maximum number of connected peers and all peers are assumed to be downloading and uploading at the same speed. Suppose there is only one uploader of the file, i.e., the first uploader who is the target of the investigation. After the first uploader publishes the file, n downloaders immediately download the file from him. Upon completing the transfer, the file is disseminated to n new users by each of the n original users. As a result, n^2 users (excluding the first uploader) obtain the file after two rounds.

The following assumptions are made for reasons of simplicity: (i) the download and upload speeds are steady and equal to 500 Kbps; (ii) the file to be transferred has a size of 10 MB; and (iii) the number of simultaneous uploads (n) for a Foxy client is five as suggested by the default maximum upload slots for Shareaza [8] (another G2 client). Figure 1 shows the hypothetical file distribution ratio based on these parameters. During the initial round of downloading, all five peers complete the file download in 160 seconds. Afterwards, the download time is shorter as

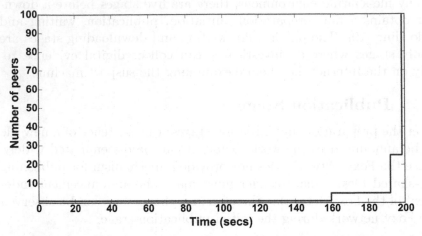

Figure 1. Number of peers possessing a file during the first 200 seconds.

more peers upload the file and the number of peers possessing the file increases exponentially. Therefore, it is extremely difficult to identify the first uploader of a shared file after the initial download burst.

6.   Proposed Investigative Process

Table 3 summarizes the experimental results and the hypothetical behavior of the Foxy network. According to our findings, leaf nodes receive filtered (O_7) and frequently-issued queries (O_8, O_9) from connected ultrapeers. New and popular keyword searches dominate the queries received by clients for similar hash-matched files (O_{10}, O_{11}).

Publication of file keywords for the Foxy network is usually performed external to the network (H_1). Therefore, the publication of new files by the first uploader is rarely detected. However, shortly after the keywords are published, the first batch of downloaders search for and download the file from the first uploader before the file is broadcasted to other users. Based on Hypotheses H_2 and H_3, the search and download processes initiate two bursts of network traffic: search query traffic inside the Foxy network and file transfer traffic outside the Foxy network.

Digital forensic investigations of suspected uploaders in the Foxy network can only be conducted after identifying their IP addresses. According to H_3, the Foxy network has to be monitored and packets have to be captured before the download burst. As long as the IP address of the uploading peer is identified before the file is widely broadcast, the chances of identifying the first uploader are comparatively higher.

Table 3. Summary of experimental findings and hypothesized behavior.

Label	Description
O_1	Foxy clients connect to ultrapeer nodes provided by the GWC server.
O_2	Each leaf node randomly connects to no more than 32 ultrapeer nodes.
O_3	Each ultrapeer connects to no more than 200 leaf nodes.
O_4	For each reconnection session, a leaf node connects to different sets of ultrapeers.
O_5	Newly-shared file with an uncommon file name can be difficult to find in the file searching process.
O_6	Set of query hits returned is affected by connected and neighboring ultrapeers.
O_7	Queries from leaf nodes are regulated and filtered at ultrapeer nodes.
O_8	Leaf nodes without any shared files also receive queries from neighboring ultrapeers.
O_9	Plaintext queries and query hash values are received at leaf nodes.
O_{10}	$Q2$ packets can arrive at leaf nodes even if they do not possess the file for sharing.
O_{11}	$Q2$ packets with the relevant hash values dominate the Foxy network shortly after an attractive keyword is announced.
O_{12}	Percentage of the frequently-asked query gradually decreases after the majority of Foxy users have downloaded the file.
H_1	No observable changes are induced during or shortly after the announcement of a keyword.
H_2	Number of $Q2$ packets querying a keyword increases after the announcement of a keyword.
H_3	Number of peers possessing the newly-shared file increases exponentially after the announcement of a popular keyword.

Monitoring search query traffic is much more useful than attempting to monitor the downloading of a file that is the subject of an investigation. File transfer occurs via a direct connection between peers; therefore, only the peers involved in the file transfer can identify the IP address of the uploader. On the other hand, query packets ($Q2$) and query hit packets ($QH2$) in search query traffic contain the requester's IP address, ultrapeer's IP address, query hash value or plaintext query keywords, real name of the file, and most importantly, the IP address from where the file can be downloaded.

Figure 2 provides an overview of the simplified Foxy investigation model. In a normal situation (Figure 2(a)), queries received by Foxy clients have different patterns. During a burst (Figure 2(b)), the same queries are received, originating from different leaf nodes. The monitoring nodes use the same query and submit it back to the Foxy network (Figure 2(c)). When the query arrives at the uploader, it returns the full name of the file and its IP address in the QH2 packet (Figure 2(d)).

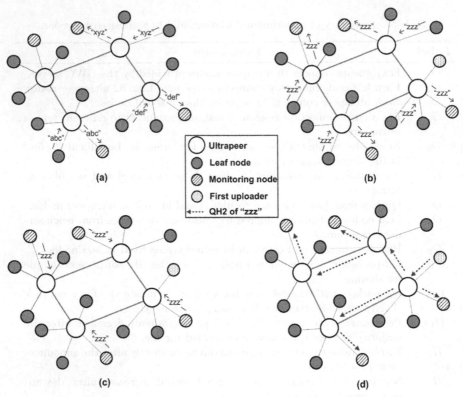

Figure 2. Simplified Foxy investigation model.

This enables us to specify the following investigative process using a customized client:

- Connect to the Foxy network and collect all $Q2$ packets from the ultrapeers.

- Re-pack and submit the identified query when a large number of $Q2$ packets querying the same keyword are observed.

- Extract the source IP address and the matched file name from all collected $QH2$ packets for analysis.

- Create a list of source IP addresses and determine the frequency of the returned IP addresses. The shorter the period between the initial identification of the burst and the generation of the list, the greater the likelihood of locating the first uploader.

7.   Conclusions

Identifying the IP addresses of P2P clients is crucial to P2P network investigations. However, it is extremely difficult to identify the first uploaders of files in the Foxy network. The investigative methodology presented in this paper leverages query packets and IP addresses to help identify uploaders in the Foxy network. The strategy is derived from observations and findings based on captured Foxy network packets. Because the Foxy network uses the G2 network protocol with minor modifications, the strategy should be applicable to general G2 network investigations.

Building on this work, we are customizing a Foxy client to analyze search behavior in an isolated experimental environment. Actual and simulated results of G2 network searches and file downloads will be collected to refine the investigative process. In addition, we plan to extend the investigative process to other "search-based" P2P networks.

References

[1] Architecture Technology Corporation, P2P Marshal Digital Forensics Software, Eden Prairie, Minnesota (p2pmarshal.atc-nycorp .com).

[2] E. Bangeman, P2P responsible for as much as 90 percent of all 'Net traffic, *Ars Technica*, September 3, 2007.

[3] BBC News, BitTorrent user guilty of piracy (news.bbc.co.uk/1/hi /technology/4374222.stm), October 25, 2005.

[4] J. Cheng, Sandvine: Close to half of all bandwidth sucked up by P2P, *Ars Technica*, June 23, 2008.

[5] M. Chesterton, Edison Chen and 7 HK stars involved in sex photos scandal, *eNews 2.0*, February 21, 2008.

[6] K. Chow, K. Cheng, L. Man, P. Lai, L. Hui, C. Chong, K. Pun, W. Tsang, H. Chan and S. Yiu, BTM - An automated rule-based BT monitoring system for piracy detection, *Proceedings of the Second International Conference on Internet Monitoring and Protection*, p. 2, 2007.

[7] K. Chow, R. Ieong, M. Kwan, P. Lai, F. Law, H. Tse and K. Tse, Security Analysis of the Foxy Peer-to-Peer File-Sharing Tool, Technical Report TR-2008-09, Department of Computer Science, University of Hong Kong, Hong Kong, 2008.

[8] Discordia, Shareaza, New York (www.shareaza.com).

[9] Gnutella2, Gnutella2 Developer Network (g2.trillinux.org).

[10] R. Ieong, P. Lai, K. Chow, M. Kwan, F. Law, H. Tse and K. Tse, Forensic investigation and analysis of peer-to-peer file-sharing networks (submitted for publication), 2009.

[11] P. Moy, Warning over rape clips, *The Standard*, Hong Kong, September 12, 2008.

[12] P. Moy and N. Patel, Covert cops hit by leaks, *The Standard*, Hong Kong, May 27, 2008.

[13] O. Nasraoui, D. Keeling, A. Elmaghraby, G. Higgins and M. Losavio, Node-based probing and monitoring to investigate the use of peer-to-peer technologies for the distribution of contraband material, *Proceedings of the Third International Workshop on Systematic Approaches to Digital Forensic Engineering*, pp. 135–140, 2008.

[14] Vastel Technology, Foxy, Hong Kong (www.gofoxy.net).

[15] Wireshark Foundation, Wireshark, San Jose, California (www.wire shark.org).

[16] Zemerick Software, Spear Forensics Software, Oak Hill, West Virginia (www.spearforensics.com/products/forensicp2p/index.aspx).

Chapter 14

DETECTING FRAUD IN INTERNET AUCTION SYSTEMS

Yanlin Peng, Linfeng Zhang and Yong Guan

Abstract   Fraud compromises the thriving Internet auction market. Studies have shown that fraudsters often manipulate their reputations through sophisticated collusions with accomplices, enabling them to defeat the reputation-based feedback systems that are used to combat fraud. This paper presents an algorithm that can identify colluding fraudsters in real time. Experiments with eBay transaction data show that the algorithm has low false negative and false positive rates. Furthermore, the algorithm can identify fraudsters who are innocent at the time of the data collection, but engage in fraudulent transactions soon after they accumulate good feedback ratings.

Keywords: Internet auctions, fraud detection

1.   Introduction

Internet auctions are a major online business. eBay, the leading Internet auction company, had a net revenue of $1.89 billion during the third quarter of 2007; it currently has a community of more than 212 million users around the world. Internet auction fraud can significantly affect this multi-billion-dollar worldwide market. In 2006, the Internet Crime Complaint Center (IC3) reported that Internet auction fraud accounted for 44.9% of all online fraud, and that the average financial loss per case was $602.50 [9].

To prevent potential fraud and encourage honest transactions, Internet auction companies have adopted reputation-based feedback systems. In these systems, users have publicly-viewable feedback ratings, and users may enter comments about each other after completing transactions. A user's feedback score is computed as the number of unique users who have left positive ratings minus the number of unique users

G. Peterson and S. Shenoi (Eds.): Advances in Digital Forensics V, IFIP AICT 306, pp. 187–198, 2009.

who have left negative ratings. Ideally, fraudsters would have low feed-back scores and/or low percentages of positive feedback ratings, discouraging honest users from participating in transactions with them.

Feedback systems can be manipulated by fraudsters in a variety of ways [3]. The assumption that an honest reputation implies honest behavior in the future is not always valid. Fraudsters often earn good reputations by making several small sales and then make fraudulent transactions on high-priced items. A more sophisticated technique involves collusions with accomplices on "virtual" transactions involving expensive items. After earning a good reputation with 50 or more positive feedback scores, a fraudster can engage in fraudulent transactions with expensive products such as computer equipment. Several researchers (see, e.g., [1, 8]) have attempted to build stronger reputation systems, but these systems are not very effective.

A complementary approach, implemented by eBay's Risk Management Group, is to manually search for fraudulent transactions. However, it is infeasible to investigate every transaction or even a large proportion of the millions of daily transactions on eBay. Some researchers [2, 5] have proposed automated methods that analyze history data to detect abnormal buying and selling behavior. Another strategy [6, 11] is to use belief propagation to identify colluding fraudsters. However, existing approaches are limited or are incapable of detecting sophisticated and hidden relationships between fraudsters and accomplices.

This paper presents an algorithm that identifies – and even predicts – colluding fraudsters in Internet auction systems. When supplied with real eBay data, the algorithm detected twenty fraudsters and predicted ten users as potential fraudsters. The algorithm engages a sliding window to deal with the fact that fraudsters have short lifetimes. The sliding window significantly reduces the computational complexity and makes it possible to accurately process large volumes of transactions in real time. Experiments with synthetic data indicate that algorithm produces very few incorrect identifications.

2.   Related Work

Approaches that engage data mining to detect fraud in Internet auction systems fall into two categories: those that detect abnormal patterns of individual users and those that detect sophisticated transaction relationships between users. Bhargava and co-workers [2] have proposed a technique that identifies auction fraud by detecting abnormal profiles and user behavior, building patterns from exposed fraudsters and discovering malicious intentions. Chau and colleagues [5] have developed a

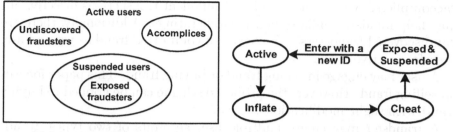

(a) Internet auction community. (b) State transitions of fraudsters.

Figure 1. Internet auction community and state transitions of fraudsters.

data mining method that generates a decision tree based on sixteen features extracted from user profiles and individual transaction histories; the decision tree is used to classify users as "legitimate" or "fraudulent." A more sophisticated method for detecting abnormal transaction relationships between users [6] uses belief propagation to discover abnormal transaction patterns between colluding fraudsters and accomplices (modeled as bipartite subgraphs in a undirected transaction graph). An improved technique [11] uses incremental belief propagation on a smaller subgraph (three-hop neighborhood) for each new transaction. However, all these methods are unable to detect collusion in Internet auction systems efficiently and accurately.

The problem addressed in this paper is similar to the dense subgraph detection problem in web graphs. Two algorithms for detecting these graphs are "trawling" [10], which enumerates all the complete bipartite subgraphs; and "shingling" [7], which extracts dense subgraphs. However, these algorithms do not address the problem presented in this paper in an efficient manner.

3. Fraud Model and Problem Definition

This section discusses the fraud model underlying Internet auction systems and defines the fundamental problem addressed in this paper.

3.1 Fraud Model

Figure 1(a) shows a classification of user accounts (also called "users") in an Internet auction community. Every person who registers successfully receives an "active" user account. Some registered users may be suspended and cannot conduct transactions for various reasons (e.g., for committing fraudulent transactions). A "Not-A-Registered-User" message is shown on the profile page of suspended users. Exposed fraudsters are placed in the category of suspended users. Some users are "accomplices," who do not conduct fraudulent transactions directly, but may help fraudsters inflate their positive ratings. Additionally, there are "undiscovered fraudsters" who have not committed fraud as yet and are, therefore, active users.

A user may engage in selling fraud or buying fraud. This paper focuses on selling fraud. However, the proposed scheme can be applied to buying fraud with slight modifications.

A fraudster may create multiple user accounts of two types, fraudster accounts and accomplice accounts. Fraudster accounts are used to commit fraud after good reputations have been earned. A fraudster's account may be suspended after a fraudulent transaction, but by then the fraudster may have already made a profit and could return as a new user. Figure 1(b) shows the state transitions made during the lifetime of an auction fraudster. Note that the lifetime of a user account ranges from the time it is created and registered to the time it is suspended. The accomplice accounts, which cannot be identified by current detection methods, remain as legitimate, active users. Accomplices typically serve multiple fraudsters. Therefore, the relationships between fraudsters and accomplices can be expressed as "bipartite cores" (i.e., complete bipartite subgraphs) in a transaction graph.

3.2   Problem Definition

We model the relationships between users in an Internet auction system as a directed transaction graph and colluding patterns as bipartite cores. A bipartite core is a complete bipartite subgraph consisting of two groups of nodes A and B. Every node in A is connected to all the nodes in B. Detecting colluding fraudsters from transaction data is equivalent to detecting bipartite cores with size $\geq s \times t$ in the transaction graph. A transaction graph is a directed graph $G = (V, E)$, where V is the set of nodes representing users and E is the set of directed edges representing transactions between two users. A bipartite core consists of two sets of nodes, "parent nodes" (sellers) and "child nodes" (buyers). Each edge is a transaction with a timestamp T that denotes when the transaction occurred. Edges are added to the transaction graph in chronological order.

One method for extracting bipartite cores is to process the entire transaction graph. However, based on our analysis of eBay data, fraudsters usually have short lifetimes because they do not wait long before committing fraud. Once they commit fraud, they are quickly reported and suspended. Consequently, detecting fraudsters only requires the extraction of the bipartite cores from a subgraph of the transaction graph.

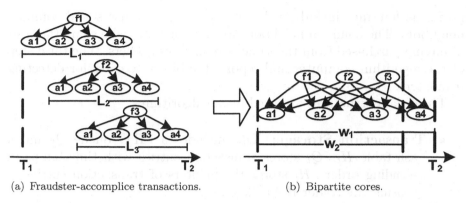

(a) Fraudster-accomplice transactions.   (b) Bipartite cores.

Figure 2.   Sliding window model and bipartite cores within windows.

In the example in Figure 2, we assume that a fraudster f_i ($i = 1, 2, 3$) has a lifetime L_i. During their lifetimes, fraudsters engage in transactions with accomplices a_j ($j = 1, 2, 3, 4$). The earliest transaction between them occurs at time T_1 and the latest at time T_2 ($T_1 < T_2$). If the bipartite core is extracted from a subgraph of transactions bound by a window (or time interval) $w_1 \geq (T_2 - T_1)$, the collusion forms a 3×4 bipartite core (Figure 2(b)). If a smaller window of size w_2 is chosen, the subgraph is smaller and only some of the fraudsters' transactions are included; the resulting bipartite core is 3×3. Based on our analysis of eBay data, a three-month window size is sufficient to identify most fraudsters.

However, not all bipartite cores indicate collusion. Sometimes, small bipartite cores are normal patterns among honest users, especially when the bipartite cores exist in the neighborhood of power users who conduct large numbers of transactions with their customers. Consequently, our algorithm excludes power users and small bipartite cores in order to capture fraud patterns more accurately.

In summary, this paper focuses on the problem of detecting all bipartite cores with size $\geq s \times t$ that represent collusion relationships between fraudsters and accomplices from a large dynamically-changing transaction graph G.

4.   Fraud Detection Algorithm

In the preliminary step of the algorithm, non-positive feedback ratings are removed because fraudsters and accomplices always leave positive ratings for each other. The detection process then involves three steps. In the first step, transactions that are outside the sliding window with respect to the newly arrived transaction are removed. The second step performs filtering, including the removal of power users and common neighbors. The common neighbor filter retains users who share the trait of having purchased from the same (two or more) sellers. The third step of the algorithm computes and reports the bipartite cores for detecting fraudsters.

Three data structures are used in the algorithm:

- **Transaction Storage:** This includes a FIFO queue Q_t and a hash table H_t. Q_t stores transaction entries with timestamps in ascending order. H_t stores the pointers of transaction entries in Q_t using the trader IDs as keys.

- **Common Neighbor Counter Storage:** This hash table H_c stores the number of common neighbors. The common neighbors for a pair of parent nodes (i, k) is the intersection of their child nodes $N(i) \cap N(k)$, where $N(\cdot)$ is the set of neighboring child nodes. H_c stores $\langle i, k, C_{i,k} \rangle$ using (i, k) as keys (where $C_{i,k} = |N(i) \cap N(k)|$).

- **Bipartite Core Storage:** This hash table H_b stores the detected maximum bipartite cores using the IDs of nodes in the bipartite core as keys. A maximum bipartite core is a core that is not a subset of another bipartite core. Let $G_p(b)$ denote the group of parent nodes in a bipartite core b and $G_c(b)$ denote the group of child nodes. For any two bipartite cores b_1 and b_2, if $G_p(b_1) \subseteq G_p(b_2)$ and $G_c(b_1) \subseteq G_c(b_2)$, only b_2 is stored; otherwise, both bipartite cores are stored.

4.1   Filtering

Two filtering functions are used to efficiently discard edges that do not contribute to qualified bipartite cores. The Power-User-Filter(i, j, R) function removes users whose reputations exceed R. Fraudsters rarely spend the time to earn extremely high reputations as lower reputations suffice for their purposes. The Common-Neighbor-Filter(i, j, t) function checks for at least t common neighbors (buyers) for a pair of parents (sellers) prior to building $s \times t$ bipartite cores. For each transaction (i, j) and parent pair (i, k) (where j is also a child of k), the child j is added to the set $N(k)$ and the common neighbor counter for k is incremented. If the maximum common neighbor counter for a parent node is at least t, the filter returns "pass" and the process enters the next step to compute the bipartite cores.

Algorithm 1 Compute-Bipartite-Cores

---

input: new transaction (i, j), detection size t

1: **for all** k such that parent pair $(i, k) \in H_c$ **do**
2:    **if** $C_{i,k}$ is increased to t after adding (i, j) **then**
3:       new bipartite core $b \leftarrow \{\{i, k\}, N(i) \cap N(k)\}$
4:       **if** $b \notin H_b$ **then**
5:          $Insert(H_b, node, b)$
6:       **end if**
7:    **end if**
8: **end for**
9: **for all** bipartite core $\widehat{b} \in H_b$ **do**
10:    **if** $|G_p(\widehat{b}) \cap N(j)| \geq 2$ **then**
11:       new bipartite core $b \leftarrow \{G_p(\widehat{b}) \cap N(j), G_c(\widehat{b}) \cup \{j\}\}$
12:    **else if** $|G_c(\widehat{b}) \cap N(i)| \geq t$ **then**
13:       new bipartite core $b \leftarrow \{G_p(\widehat{b}) \cup \{i\}, G_c(\widehat{b}) \cap N(i)\}$
14:    **end if**
15:    **if** $b \notin H_b$ **then**
16:       $Insert(H_b, node, b)$
17:    **end if**
18: **end for**

4.2   Computing Bipartite Cores

The process of computing bipartite cores builds larger bipartite cores from smaller ones. As shown in Algorithm 1, the smallest $2 \times t$ bipartite cores are constructed by intersecting the parent pairs whose common neighbor counters are no less than t (Lines 1–8). Then, the bipartite cores containing i or j are retrieved to check if they can be enlarged. For example, if a bipartite core b contains node i, then the intersection of $G_p(b)$ and $N(j)$ is checked. If the size of the intersection is large enough, the existing bipartite core b can be enlarged by adding the child node j. If an enlarged bipartite core is not a subset of an existing core, it is added to the bipartite core storage.

4.3   Reporting

The reporting function lists the users present in "fraudulent bipartite cores." A fraudulent bipartite core is a bipartite core with size $\geq s \times t$ that contains at least one exposed fraudster. We believe that users in these bipartite cores will commit fraud with a high probability. Non-fraudulent bipartite cores are also stored. As soon as a user is identified as a fraudster, all the other users in the bipartite core are reported.

5.   Analysis of Countermeasures

Armed with the details of the algorithm, fraudsters can implement two countermeasures to evade detection. In the following discussion, we assume that: (i) n accomplices support as many fraudsters as possible and evade detection; (ii) a fraudster needs to earn a reputation of at least r before committing fraud; and (iii) the detection size is (s,t) and $t \leq r \leq n$.

The first countermeasure is to form bipartite cores of size $i \times j$ where $i \geq s, j < t$. Thus, less than t accomplices support at least s fraudsters, which is not detected. However, a fraudster cannot earn a reputation of r from one group of j accomplices within a sliding window. The fraudster must wait longer than the time period covered by one sliding window to earn the required reputation or he must collude with additional accomplices. Even if the fraudster waits longer to earn a reputation of r, the system administrator can enlarge the sliding window to counter this strategy. If the fraudster attempts to earn the desired reputation within one sliding window, a subset of the i fraudsters must collude with other accomplices, forming bipartite cores with size $f \times g$ where $f < s, g \geq r$. This is the same as the second countermeasure.

The second countermeasure is to form bipartite cores of size $i \times j$ where $i < s, j \geq r$. Thus, at least r accomplices support fewer than s fraudsters. Such collusion is not detected and a fraudster can earn a reputation of r within a short time. However, the proposed algorithm can still set an upper bound on the number of supported fraudsters in this case. To maximize the number of supported fraudsters, it is necessary to maximize the number of j-accomplice groups, maximize the number of fraudsters supported by each accomplice group, and support different fraudsters for each accomplice group. Assume that there are m ways to choose accomplice groups such that each group has at least r accomplices and every two groups have intersections of at most $(t-1)$. According to the non-uniform Ray-Chaudhuri-Wilson inequality, $m \leq \binom{n}{t-1} + \cdots + \binom{n}{0}$. Thus, at most $m \cdot (s-1)$ fraudsters can be supported by a group of n accomplices without being detected. To address this countermeasure, the systems administrator can choose small values of s and t to reduce the number of fraudsters that can be supported by an accomplice group.

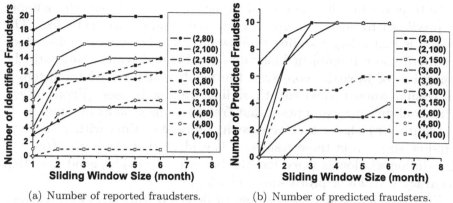

(a) Number of reported fraudsters.   (b) Number of predicted fraudsters.

Figure 3. Results obtained with eBay data.

6.   Experimental Results

This section describes the experimental results obtained using real eBay data and synthetic data.

6.1   Results with eBay Data

The first set of experiments used eBay data collected from September 3, 2007 to November 3, 2007. The data was gathered by a crawling program that started with a list of eBay user IDs and obtained their feedback profiles in a breadth-first manner. In all, 5,795,314 transactions and 3,406,783 eBay users were crawled. Of these users, 43 were fraudsters who received many negative ratings and were suspended. Since the identified fraudsters may not have colluded with accomplices, our algorithm may only identify some of the fraudsters.

The algorithm was executed on a 2 GB dual core Mac Pro. Transactions were input in ascending order of transaction timestamps. The power user threshold was set to 3000. Several detection sizes and sliding window sizes were tested in our experiments.

Figure 3(a) shows that as many as twenty of the identified fraudsters were reported by the algorithm. Figure 3(b) shows the most important result – up to 10 users were predicted to be fraudsters. These users were not identified as fraudsters at the time the data was collected. After the users were reported by the algorithm, we went back to check them out and discovered that they had received many negative ratings and had been suspended. If our algorithm had been applied in a real-world setting, these users would have been flagged before they committed fraud.

Note also that the number of reported fraudsters and the number of predicted fraudsters increase when the window size increases. The reason is that a larger sliding window produces larger subgraphs, which results in more fraudulent bipartite cores being detected. Interestingly, regardless of bipartite core size, the number of detected fraudsters is maximum when a three-month sliding window is used (Figure 3(a)). This confirms our assumption that fraudsters have short lifetimes.

Figure 4(a) shows the increase in processing time with respect to window size. From these results, it is evident that the best setting for the sliding window is three months because it provides good detection accuracy with lower processing overhead.

Figure 3 shows that the number of detected fraudsters also depends on the bipartite core detection size. For larger sizes, fewer fraudsters are identified because fraudsters in small collusions are not detected. The detection sizes (2, 80) and (2, 100) identify the most fraudsters. Also, the processing time increases for larger detection sizes. Based on these results, we recommend that 2×100 detection sizes be used for fraud detection.

6.2   Results with Synthetic Data

Synthetic data was used to evaluate the false positive and false negative rates for the algorithm. The false negative rate is defined as the number of distinct nodes in the injected bipartite cores that are not reported divided by the number of total distinct nodes in all the injected bipartite cores. The false positive rate is defined as the number of distinct nodes not in the injected bipartite cores but that are reported divided by the number of total distinct nodes not in any injected bipartite core.

Since the distribution of eBay feedback obeys a power law distribution [12], R-MAT [4] was used to generate a random power law transaction graph of about 100,000 nodes with an average degree of 1.7. Next, ten bipartite cores of known sizes were injected as fraudster bipartite cores as follows: (i) the sizes of the two groups were randomly chosen to be between 3 and 10; (iii) nodes in each group were randomly chosen among all nodes; and (iii) a fraud lifetime of 10,000 units was defined to represent a short lifetime. The transaction timestamps in each fraudulent bipartite core were chosen randomly within the fraudster lifetime. Some nodes in these bipartite cores were randomly selected as identified fraudsters. Then, the transaction graph was input to the algorithm and the fraudsters were reported.

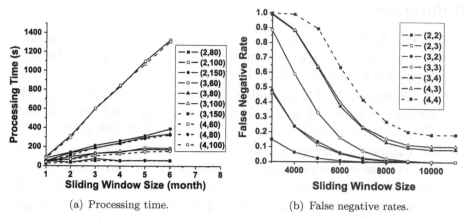

(a) Processing time.   (b) False negative rates.

Figure 4.   Processing time (eBay data) and false negative rates (synthetic data).

Figure 4(b) shows the average results over 500 executions. Note that the false negative rate decreases when the window size increases. A larger window ensures larger subgraphs and more fraudulent cores; this reduces the likelihood of missing fraudulent nodes. When the window size is equal to the fraud lifetime, the false negative rate reaches its lowest value for each detection size. Note also that the false negative rate decreases when the detection size decreases. Most of the fraudulent nodes are detected even when a very small window size of 0.4 times the fraud lifetime and a detection size of 2×2 are used.

The false positive rates are close to zero for all window sizes and detection sizes. Honest nodes are wrongly reported only if they happen to be in a fraudulent core, which has a very small probability. Thus, our algorithm can distinguish honest users from fraudsters very effectively. Moreover, the sliding window method reduces not only the processing overhead but also the false positive rate.

7.   Conclusions

Fraudulent activities can compromise the multi-billion-dollar Internet auction market. Existing solutions are either limited or are incapable of detecting colluding fraudsters. The algorithm presented in this paper can detect colluding fraudsters at runtime with good accuracy. Experiments with real and synthetic data demonstrate that the algorithm can detect fraudsters and, more importantly, predict potential fraudsters.

References

[1] S. Ba, A. Whinston and H. Zhang, Building trust in online auction markets through an economic incentive mechanism, *Decision Support Systems*, vol. 35(3), pp. 273–286, 2003.

[2] B. Bhargava, Y. Zhong and Y. Lu, Fraud formalization and detection, *Proceedings of the Fifth International Conference on Data Warehousing and Knowledge Discovery*, pp. 330–339, 2003.

[3] M. Calkins, My reputation always had more fun than me: The failure of eBay's feedback model to effectively prevent online auction fraud, *Richmond Journal of Law and Technology*, vol. VII(4), 2001.

[4] D. Chakrabarti, Y. Zhan and C. Faloutsos, R-MAT: A recursive model for graph mining, *Proceedings of the SIAM International Conference on Data Mining*, 2004.

[5] D. Chau and C. Faloutsos, Fraud detection in electronic auctions, *Proceedings of the European Web Mining Forum*, 2005.

[6] D. Chau, S. Pandit and C. Faloutsos, Detecting fraudulent personalities in networks of online auctioneers, *Proceedings of the Tenth European Conference on Principles and Practice of Knowledge Discovery in Databases*, pp. 103–114, 2006.

[7] D. Gibson, R. Kumar and A. Tomkins, Discovering large dense subgraphs in massive graphs, *Proceedings of the Thirty-First International Conference on Very Large Data Bases*, pp. 721–732, 2005.

[8] D. Houser and J. Wodders, Reputation in auctions: Theory and evidence from eBay, *Journal of Economics and Management Strategy*, vol. 15(2), pp. 353–369, 2006.

[9] Internet Crime Complaint Center, Internet Crime Report 2006 (www.ic3.gov/media/annualreport/2006_IC3Report.pdf), 2006.

[10] R. Kumar, P. Raghavan, S. Rajagopalan and A. Tomkins, Trawling the web for cyber communities, *Computer Networks*, vol. 31(11-16), pp. 1481–1493, 1999.

[11] S. Pandit, D. Chau, S. Wang and C. Faloutsos, Netprobe: A fast and scalable system for fraud detection in online auction networks, *Proceedings of the Sixteenth International Conference on the World Wide Web*, pp. 201–210, 2007.

[12] M. Zhou and F. Hwang, PowerTrust: A robust and scalable reputation system for trusted peer-to-peer computing, *IEEE Transactions on Parallel and Distributed Systems*, vol. 18(4), pp. 460–473, 2007.

V

FORENSIC COMPUTING

Chapter 15

A CLOUD COMPUTING PLATFORM FOR LARGE-SCALE FORENSIC COMPUTING

Vassil Roussev, Liqiang Wang, Golden Richard and Lodovico Marziale

Abstract   The timely processing of massive digital forensic collections demands the use of large-scale distributed computing resources and the flexibility to customize the processing performed on the collections. This paper describes MPI MapReduce (MMR), an open implementation of the MapReduce processing model that outperforms traditional forensic computing techniques. MMR provides linear scaling for CPU-intensive processing and super-linear scaling for indexing-related workloads.

Keywords: Cluster computing, large-scale forensics, MapReduce

1.   Introduction

According to FBI statistics [4], the size of the average digital forensic case is growing at the rate of 35% per year – from 83 GB in 2003 to 277 GB in 2007. With storage capacity growth outpacing bandwidth and latency improvements [9], forensic collections are not only getting bigger, but are also growing significantly larger relative to the ability to process them in a timely manner. There is an urgent need to develop scalable forensic computing solutions that can match the explosive growth in the size of forensic collections.

The problem of scale is certainly not unique to digital forensics, but forensic researchers have been relatively slow to recognize and address the problem. In general, three approaches are available to increase the processing performed in a fixed amount of time: (i) improving algorithms and tools, (ii) using additional hardware resources, and (iii) facilitating human collaboration. These approaches are mutually independent and support large-scale forensics in complementary ways. The first approach supports the efficient use of machine resources; the second permits additional machine resources to be deployed; the third leverages human

G. Peterson and S. Shenoi (Eds.): Advances in Digital Forensics V, IFIP AICT 306, pp. 201–214, 2009.
© IFIP International Federation for Information Processing 2009

expertise in problem solving. In order to cope with future forensic collections, next generation forensic tools will have to incorporate all three approaches. This paper focuses on the second approach – supporting the use of commodity distributed computational resources to speedup forensic investigations.

Current forensic tools generally perform processing functions such as hashing, indexing and feature extraction in a serial manner. The result is that processing time grows as a function of the size of the forensic collection. An attractive long-term solution is to deploy additional computational resources and perform forensic processing in parallel. A parallel approach is more sustainable because storage capacities and CPU processing capabilities increase at approximately the same rate as predicted by Moore's law and as observed by Patterson [9]. In other words, as the average collection size doubles, doubling the amount of computational resources maintains the cost of the expansion constant over time. Furthermore, the widespread adoption of data center technologies is making the logistic and economic aspects of the expansion of computational resources even more favorable.

The data center approach is clearly a long-term goal. The first step is to leverage existing computational resources in a forensic laboratory. For example, a group of hosts in a local area network could be temporarily organized as an *ad hoc* cluster for overnight processing of large forensic collections. The main impediment is the lack of a software infrastructure that enables forensic processing to be scaled seamlessly to the available computing resources. This paper describes a proof-of-concept software infrastructure that could make this vision a reality.

2.   Related Work

Early applications of distributed computing in digital forensics demonstrated that it is possible to achieve linear speedup (i.e., speedup proportional to the number of processors/cores) on typical forensic functions [11]. Furthermore, for memory-constrained functions, it is possible to achieve super-linear speedup due to the fact that a larger fraction of the data can be cached in memory.

Several efforts have leveraged distributed computing to address digital forensic problems. ForNet [12] is a well-known project in the area of distributed forensics, which focuses on the distributed collection and querying of network evidence. Marziale and co-workers [7] have leveraged the hardware and software capabilities of graphics processing units (GPUs) for general-purpose computing. Their approach has the same goals as this work and is, in fact, complementary to it. Clearly, using

CPU and GPU resources on multiple machines will contribute to more speedup than using CPUs alone.

Recently, AccessData started including support for multi-core processors under its FTK license. It currently offers limited distributed capabilities for specialized tools (e.g., for password cracking). FTK also includes a database back-end that can manage terabytes of data and utilizes CPU and memory resources to perform core forensic functions.

3.   MapReduce

MapReduce [3] is a distributed programming paradigm developed by Google for creating scalable, massively-parallel applications that process terabytes of data using large commodity clusters. Programs written using the MapReduce paradigm can be automatically executed in parallel on clusters of varying size. I/O operations, distribution, replication, synchronization, remote communication, scheduling and fault tolerance are performed without input from the programmer, who is freed to focus on application logic.

After the early success of MapReduce, Google used the paradigm to implement all of its search-related functions. This is significant because of Google's emphasis on information retrieval, which is also at the heart of most forensic processing. Conceptually, information retrieval involves the application of a set of n functions (parsing, string searching, calculating statistics, etc.) to a set of m objects (files). This yields $n \times m$ tasks, which tend to have few, if any, dependencies and can, therefore, be readily executed in parallel.

Phoenix [10] is an open-source prototype that demonstrates the viability of the MapReduce model for shared memory multi-processor/multi-core systems. It provides close to linear speedup for workloads that are relevant to forensic applications (e.g., word counts, reverse indexing and string searches). The main limitation is that it executes on a single machine and has no facilities to scale it to cluster environments.

Hadoop [1] is an open-source Java implementation of the MapReduce model that has been adopted as a foundational technology by large Internet companies such as Yahoo! and Amazon. The National Science Foundation has partnered with Google and IBM to create the Cluster Exploratory (CluE), a cluster of 1,600 processors that enables scientists to create large-scale applications using Hadoop. One concern about the Java-based Hadoop platform is that it is not as efficient as Google's C-based platform. This may not be a significant issue for large deployments, but it can impact efficiency when attempting to utilize relatively small clusters. Another concern is that Hadoop's implementation requires the deployment of the Hadoop File System (HDFS), which is implemented as an abstraction layer on top of the existing file system. This reduces I/O efficiency and complicates access to raw forensic images.

4.   MPI MapReduce

MapReduce is a powerful conceptual model for describing typical forensic processing. However, the Hadoop implementation is not efficient enough for deployment in a forensic laboratory environment. To address this issue, we have developed MPI MapReduce (MMR) and use it to demonstrate that the basic building blocks of many forensic tools can be efficiently realized using the MapReduce framework. Note however that an actual tool for use in a forensic laboratory environment would require additional implementation effort.

Our MMR implementation leverages two technologies, the Phoenix shared-memory implementation of MapReduce [10] and the Message Passing Interface (MPI) distributed communication standard [8]. Specifically, it augments the Phoenix shared-memory implementation with MPI to enable computations to be distributed to multiple nodes.

MPI is designed for flexibility and does not prescribe any particular model of distributed computation. While this is generally an advantage, it is also a drawback because developers must possess an understanding of distributed programming and must explicitly manage distributed process communication and synchronization.

MMR addresses this drawback by providing a middleware platform that hides the implementation details of MPI, enabling programmers to focus on application logic and not worry about scaling up computations. Additionally, MMR automatically manages communication and synchronization across tasks. All this is possible because MMR engages MapReduce as its distributed computation model.

4.1   MapReduce Model

The MapReduce computation takes a set of input key/value pairs and produces a set of output key/value pairs. The developer expresses the computation using two functions, map and reduce. Function map takes an input pair and produces a set of intermediate key/value pairs. The runtime engine then automatically groups together all the intermediate values associated with an intermediate key I and passes them to the reduce function. The reduce function accepts the intermediate key I and a set of values for the key, and uses them to produce another set of values. The I values are supplied to the reduce function via an iterator, which allows arbitrarily large lists of values to be passed.

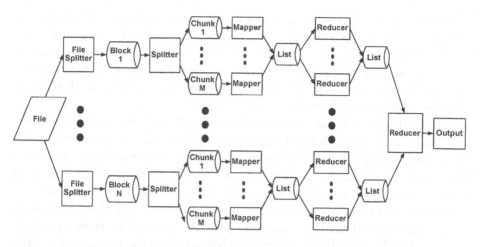

Figure 1. MMR data flow.

As an example, consider the Wordcount problem: given a set of text documents, count the number of occurrences of each word. In this case, the map function uses the word as a key to construct pairs of the form (word, 1). For n distinct words in the document, the runtime system creates n pairs and feeds them to n different instances of the reduce function. The reduce function counts the number of elements in its argument list and outputs the result.

There are several possibilities for data parallelism in this solution. In the case of the map function, it is possible to create as many independent instances as there are documents. Large documents could be split into pieces to achieve better load balance and higher levels of concurrency. The reduce computation is also parallelizable, allowing for as many distinct instances as there are distinct words. The only major parallelism constraint is that all the map instances must complete their execution before the reduce step is launched. Note that the programmer does not have to be concerned about the size of the inputs and outputs, and the distribution and synchronization of the computations. These tasks are the responsibility of the runtime environment, which works behind the scenes to allocate the available resources to specific computations.

4.2 MPI MapReduce Details

Figure 1 presents the data flow in an MMR application. A single data file is used as input to simplify the setup and isolate file system influence on execution time. A file splitter function splits the input into N equal blocks, where N is the number of available machines (nodes).

Each node reads its assigned block of data and splits the block into M chunks according to the number of mapper threads to be created at each node. M is typically set to the level of hardware-supported concurrency. This is based on the number of threads that the hardware can execute in parallel and the cache space available for the mappers to load their chunks simultaneously without interference.

After the threads are created, each thread receives a chunk of data and the programmer-defined `map` function is invoked to manipulate the data and produce key/value pairs. If the programmer has specified a reduction function, the results are grouped according to keys and, as soon as the mapper threads complete, a number of reducer threads are created to complete the computation with each reducer thread invoking the programmer-defined `reduce` function. After the reduction, each node has a reduced key/value pair list, which it sends to the master node. The master receives the data and uses a similar `reduce` function to operate on the received key/value pairs and output the final result.

Figure 2 illustrates the MMR flow of execution at each node and the basic steps to be performed by a developer. Note that the functions with an `mmr` prefix are MMR API functions supplied by the infrastructure.

The first step is to invoke the system-provided `mmrInit()` function, which performs a set of initialization steps, including MPI initialization. Next, `mmrSetup()` is invoked to specify arguments such as file name, unit size, number of map/reduce threads at each node and the list of application-defined functions (e.g., key comparison, map and reduce).

Many of the mandatory arguments have default values to simplify routine processing. By default, `Setup()` automatically opens and maps the specified files to memory (this can be overridden if the application needs access to the raw data). After the data is mapped into memory, the `splitData()` function calculates the offset and length of the data block at a node; and `setArguments()` sets all the arguments for map, reduce and MPI communication.

After the initialization steps are done, the application calls `mmr()` to launch the computation. The final result list is generated at the master node and is returned to the application. More complicated processing may require multiple map/reduce rounds. In such an instance, MMR invokes `mmrCleanup()` to reset the buffers and state information, and may setup and execute another MapReduce round using `mmrSetup()` and `mmr()`.

The `mmr()` function invokes `mmrMapReduce()` to perform map/reduce at a node. If the node is not the master, it packs the result (key/value pairs) into an MPI buffer and sends it to the master node. Since MMR does not know the data types of the keys and values, the developer must

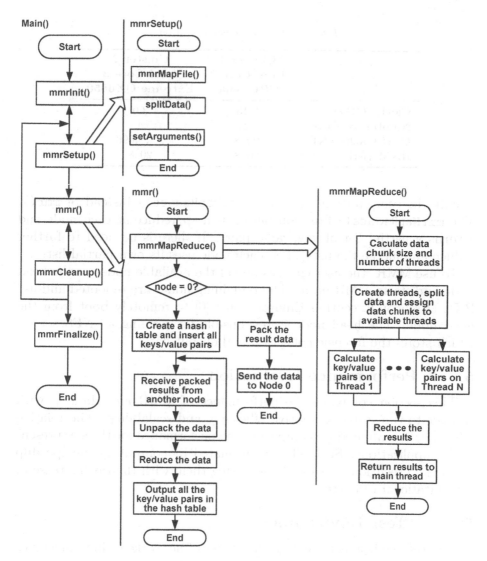

Figure 2. MMR flowchart.

write functions to pack and unpack the data. If the node is the master, after `mmrMapReduce()` is invoked, a hash table is generated to store hashed keys. The hash table is used to accelerate the later reduction of values on the same keys. The master receives byte streams from each node, unpacks them into key/value pairs, aggregates them into a list, and returns them to the nodes for the reduction step. If an application does not need each node to send its partial result to the master, it can short

Table 1.   Hardware configuration.

	Cluster 1 Intel Core 2 CPU 6400	Cluster 2 Intel Core 2 Extreme QX6850
Clock (GHz)	2.13	3.0
Number of Cores	2	4
CPU Cache (KB)	2048	2 × 4096
RAM (MB)	2048	2048

circuit the computation by calling `mmrMapReduce()` instead of `mmr()`. The `mmrMapReduce()` function performs map/reduce at each node and returns a partial list of key/value pairs. If there is no need to further reduce and merge the lists, then each node uses its own partial list.

To use MMR, the user chooses one of the available nodes as the head node and starts MMR on it. The remaining nodes are rebooted and use PXE (Preboot Execution Environment) [6] to remotely boot from the head node. The head node is responsible for running an NFS server, which stores the forensic collection and output data.

5.   Performance Evaluation

This section compares the performance of MMR and Hadoop with respect to two criteria, relative efficiency and scalability. The relative efficiency is evaluated by comparing the performance for three representative applications. Scalability is evaluated by computing the speedup vs. serial execution times and comparing them with the raw increase in computational resources.

5.1   Test Environment

The test environment included a cluster of networked Linux machines configured with a central user account management tool, `gcc 4.2`, `ssh`, `gnumake`, OpenMPI and a network file system. The experiments employed two *ad hoc* clusters of laboratory workstations, Cluster 1 and Cluster 2. Cluster 1 comprised three Dell dual-core machines while Cluster 2 comprised three Dell quad-core machines.

Table 1 presents the configurations of the machines in the clusters. Note that the quad-core machines were run with a 2 GB RAM configuration to permit the comparison of results. All the machines were configured with Ubuntu Linux 8.04 kernel version 2.6.24-19, Hadoop and MMR. The network setup included gigabit Ethernet NICs and a Cisco 3750 switch.

Note that the Hadoop installation uses HDFS whereas MMR uses NFS. HDFS separates a file into chunks and distributes them among nodes. When a file is requested, each node sends its chunks to the destination node. Wherever possible, we discounted I/O times and focused on CPU processing times. However, we found this to be somewhat difficult with Hadoop because of the lack of control over data caching in HDFS.

Our first experiment was designed to compare three Hadoop applications (wordcount, pi-estimator and grep) to their functional equivalents written in MMR. No changes were made to the Hadoop code except to add a timestamp for benchmarking purposes. The three applications are representative of the processing encountered in a typical forensic environment. The wordcount program calculates the number of occurrences (frequency) of each word in a text file. Word frequency calculations are of interest because they are the basis for many text indexing algorithms. The pi-estimator program calculates an approximation of π using the Monte Carlo estimation method; it involves a pure computational workload with almost no synchronization/communication overhead. Many image processing algorithms are CPU-bound so this test provides a baseline assessment of how well the infrastructure can utilize computational resources. The grep program searches for matched lines in a text file based on regular expression matching and returns the line number and the entire line that includes the match. It is one of the most commonly used tools in digital forensics. Note that Hadoop grep only returns the number of times the specified string appears in the file, which is weaker than the Linux grep command. MMR grep can return the line numbers and the entire lines (like the Linux grep); however, in order to permit comparisons, it only returns the counts as in the case of Hadoop.

5.2 Benchmark Execution Times

We tested wordcount and grep with 10 MB, 100 MB, 1,000 MB and 2,000 MB files, and pi-estimator with 12,000 to 1,200,000,000 points. The test results were averaged over ten runs with the total number of map/reduce threads equal to the hardware concurrency factor (note that the first and last runs are ignored). All the execution runs were performed on Cluster 2 and the results (in seconds) are shown in Figure 3 (lower values are better).

In the case of wordcount, MMR is approximately fifteen times faster than Hadoop for large (1 GB or more) files and approximately 23 times faster for small files. For grep, MMR is approximately 63 times faster

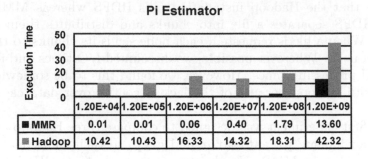

Pi Estimator	1.20E+04	1.20E+05	1.20E+06	1.20E+07	1.20E+08	1.20E+09
■ MMR	0.01	0.01	0.06	0.40	1.79	13.60
■ Hadoop	10.42	10.43	16.33	14.32	18.31	42.32

Wordcount	10 MB	100 MB	1000 MB	2000 MB
■ MMR	0.43	1.17	7.09	13.35
■ Hadoop	10	27	109	205

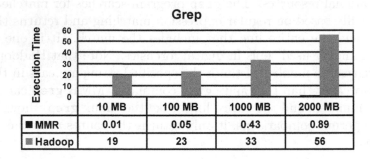

Grep	10 MB	100 MB	1000 MB	2000 MB
■ MMR	0.01	0.05	0.43	0.89
■ Hadoop	19	23	33	56

Figure 3.   Execution times for three applications.

than Hadoop in the worst case. For `pi-estimator` (which measures the purely computational workload), MMR is just over three times faster for the largest point set. Overall, it is evident that Hadoop has higher start-up costs so the times for the longer runs are considered to be more representative.

5.3   Scalability

Figure 4 compares the execution times for the three applications under MMR and Hadoop for each cluster. Since `pi-estimator` is easily

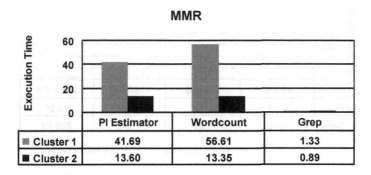

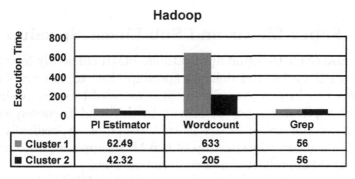

Figure 4. Benchmark times on Cluster 1 and Cluster 2.

parallelizable and scales proportionately to hardware improvements of the CPU, one would not expect caching or faster memory to make any difference. Note that the processors in Cluster 2 are newer designs compared with those in Cluster 1, and the expected raw hardware speedup is a function of the differences in the clock rates and the number of cores. Since Cluster 2 has a 50% faster clock and twice the number of cores, one would expect a speedup factor of 3 (i.e., three times faster).

The execution times for MMR show the expected improvement factor of 3 for the `pi-estimator` benchmark. In contrast, the Hadoop version only has an improvement factor of 0.5. This is a curious result for which we have no explanation, especially given the factor of 3 improvement for the `wordcount` benchmark. MMR produces a speedup factor of 4.2 in the case of `wordcount`, which exceeds the pure CPU speedup factor of 3 due to faster memory access and larger caches. In the case of `grep`, which is a memory-bound application, the speedup is dominated by faster memory access with a minor contribution from the CPU speedup. Unfortunately, we could not measure the speedup for the Hadoop version.

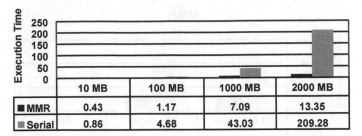

	10 MB	100 MB	1000 MB	2000 MB
■ MMR	0.43	1.17	7.09	13.35
■ Serial	0.86	4.68	43.03	209.28

Figure 5. Wordcount: MMR vs. serial execution.

5.4   Super-Linear and Sub-Linear Speedup

In the case of CPU-bound applications, MMR efficiently leverages the available CPU cycles and delivers speedup close to the raw hardware improvements. In a real-world setting, we would expect a significant number of applications not to adhere to this model; consequently, we must consider the use of MMR in such scenarios. Specifically, we use Cluster 2 to compare the speedup of two MMR applications, wordcount and bloomfilter, relative to their serial versions.

The bloomfilter application hashes a file in 4 KB blocks using SHA-1 and inserts them into a Bloom filter [2]. Then, a query file of 1 GB is hashed in the same way and the filter is queried for matches. If matches are found, the hashes triggering them are returned as the result. In the test case, the returned result is about 16 MB. To isolate and understand the effects of networking latency, we created two versions: MMR and MMR Send. MMR completes the computation and only returns a total count. On the other hand, MMR Send passes the actual hash matches to the master node.

Figure 5 compares the results of MMR and serial execution for the wordcount application. Note that wordcount scales in a super-linear fashion with 15.65 times the speedup. In fact, the relative speedup factor (speedup/concurrency) is approximately 1.3, which is close to the 1.4 value mentioned above.

Figure 6 compares the results of MMR and serial execution for the bloomfilter application. In the case of the bloomfilter application, because the computation is relatively simple and the memory access pattern is random (no cache benefits), memory and I/O latency become the bottleneck factors for speedup. In the 1 GB and 2 GB experiments, the MMR version achieves a speedup factor of 9 whereas MMR Send has a speedup factor of only 6. For longer computations, overlapping network communication with computation would reduce some of the

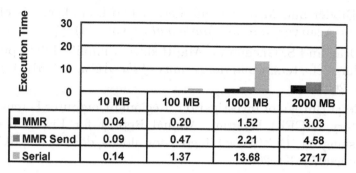

	10 MB	100 MB	1000 MB	2000 MB
■ MMR	0.04	0.20	1.52	3.03
■ MMR Send	0.09	0.47	2.21	4.58
▪ Serial	0.14	1.37	13.68	27.17

Figure 6. Bloom filter: MMR vs. serial execution.

latency, but, overall, this type of workload cannot be expected to scale as well as it does for the MMR version.

In summary, the experiments demonstrate that the proof-of-concept implementation of MMR has superior performance relative to Hadoop, the leading open-source implementation. MMR also demonstrates excellent scalability for all three types of workloads, I/O-bound, CPU-bound and memory-bound workloads.

6. Conclusions

Digital forensic tool developers need scalable development platforms that can automatically leverage distributed computing resources. The new MPI MapReduce (MMR) implementation provides this capability while outperforming Hadoop, the leading open-source solution. Unlike Hadoop, MMR efficiently and predictably scales up MapReduce computations. Specifically, for CPU-bound processing, MMR provides linear scaling with respect to the number of CPUs and CPU speed. For common indexing tasks, MMR demonstrates super-linear speedup for common indexing tasks. In the case of I/O-bound and memory-constrained tasks, the speedup is sub-linear but nevertheless substantial.

Our experimental results indicate that MMR provides an attractive platform for developing large-scale forensic processing tools. Our future research will experiment with clusters containing tens to hundreds of cores with the ultimate goal of developing tools that can deal with terabyte-size forensic collections in real time.

References

[1] Apache Software Foundation, Apache Hadoop Core, Forest Hill, Maryland (hadoop.apache.org/core).

[2] A. Broder and M. Mitzenmacher, Network applications of Bloom filters: A survey, *Internet Mathematics*, vol. 1(4), pp. 485–509, 2005.

[3] J. Dean and S. Ghemawat, MapReduce: Simplified data processing on large clusters, *Communications of the ACM*, vol. 51(1), pp. 107–113, 2008.

[4] Federal Bureau of Investigation, Regional Computer Forensics Laboratory (RCFL) Program Annual Report for Fiscal Year 2007, Washington, DC (www.rcfl.gov/downloads/documents/RCFL_Nat _Annual07.pdf), 2008.

[5] S. Ghemawat, H. Gobioff and S. Leung, The Google file system, *Proceedings of the Nineteenth ACM Symposium on Operating Systems Principles*, pp. 29–43, 2003.

[6] Intel Corporation, Preboot Execution Environment (PXE) Specification, Version 2.1, Santa Clara, California (download.intel.com /design/archives/wfm/downloads/pxespec.pdf), 1999.

[7] L. Marziale, G. Richard and V. Roussev, Massive threading: Using GPU to increase the performance of digital forensic tools, *Digital Investigation*, vol. 4(S1), pp. 73–81, 2007.

[8] Message Passing Interface Forum, MPI Forum, Bloomington, Indiana (www.mpi-forum.org).

[9] D. Patterson, Latency lags bandwidth, *Communications of the ACM*, vol. 47(10), pp. 71–75, 2004.

[10] C. Ranger, R. Raghuraman, A. Penmetsa, G. Bradski and C. Kozyrakis, Evaluating MapReduce for multi-core and multiprocessor systems, *Proceedings of the Thirteenth International Symposium on High-Performance Computer Architecture*, pp. 13–24, 2007.

[11] V. Roussev and G. Richard, Breaking the performance wall: The case for distributed digital forensics, *Proceedings of the Fourth Digital Forensic Research Workshop*, 2004.

[12] K. Shanmugasundaram, N. Memon, A. Savant and H. Bronnimann, ForNet: A distributed forensics network, *Proceedings of the Second International Workshop on Mathematical Methods, Models and Architectures for Computer Network Security*, pp. 1–16, 2003.

Chapter 16

PASSWORD CRACKING USING SONY PLAYSTATIONS

Hugo Kleinhans, Jonathan Butts and Sujeet Shenoi

Abstract   Law enforcement agencies frequently encounter encrypted digital evidence for which the cryptographic keys are unknown or unavailable. Password cracking – whether it employs brute force or sophisticated cryptanalytic techniques – requires massive computational resources. This paper evaluates the benefits of using the Sony PlayStation 3 (PS3) to crack passwords. The PS3 offers massive computational power at relatively low cost. Moreover, multiple PS3 systems can be introduced easily to expand parallel processing when additional power is needed. This paper also describes a distributed framework designed to enable law enforcement agents to crack encrypted archives and applications in an efficient and cost-effective manner.

Keywords: Password cracking, cell architecture, Sony Playstation 3

1.   Introduction

In 2006, Sebastien Boucher was arrested for possessing child pornography on his laptop while attempting to cross the Canadian–U.S. border [7]. When the FBI attempted to access his computer files at their laboratory, they discovered that the contents were password protected. The FBI requested the U.S. District Court in Vermont to order Boucher to reveal his password. However, the court ruled that Boucher was protected under the Fifth Amendment because a person cannot be compelled to "convey the contents of one's mind." According to John Miller, FBI Assistant Director of Public Affairs, "When the intent...is purely to hide evidence of a crime...there needs to be a logical and constitutionally sound way for the courts to allow law enforcement access to the evidence."

G. Peterson and S. Shenoi (Eds.): Advances in Digital Forensics V, IFIP AICT 306, pp. 215–227, 2009.

The technical solution in the Boucher case is to crack the password. However, password cracking involves advanced cryptanalytic techniques, and brute force cracking requires massive computational resources. Most law enforcement agencies neither have the technical resources nor the equipment to break passwords in a reasonable amount of time. In 2007, a major U.S. law enforcement agency began clustering computers at night in an attempt to crack passwords; these same computers were used for other purposes during the day. The approach represented an improvement over the existing situation, but it was still inadequate.

The Sony PlayStation 3 (PS3) system is an attractive platform for password cracking. The PS3 was designed for gaming, but it offers massive computational power for scientific applications at low cost. The architecture also permits the interconnection of multiple PS3 systems to enhance parallel processing for computationally-intensive problems.

This paper evaluates the potential benefits of using PS3 to crack passwords. An experimental evaluation of the PS3 is conducted along with an Intel Xeon 3.0 GHZ dual-core, dual-processor system and an AMD Phenom 2.5 GHz quad-core system. The results indicate that the PS3 and AMD systems are comparable in terms of computational power, and both outperform the Intel system. However, with respect to cost efficiency ("bang per buck") – arguably the most important metric for law enforcement agencies – the PS3, on the average, outperforms the AMD and Intel systems by factors of 3.7 and 10.1, respectively.

This paper also describes a distributed PS3-based framework for law enforcement agents in the field. The framework is designed to enable agents to crack encrypted archives and applications in an efficient and cost-effective manner.

2.   Cell Broadband Engine Architecture

Kutaragi from Sony Corporation created the cell broadband engine (CBE) architecture in 1999 [1, 2, 12]. His design, inspired by cells in a biological system, led to a large-scale collaborative effort between Sony, Toshiba and IBM [4]. The efforts focused on developing a powerful, cost-effective architecture for video game consoles. Although the CBE was designed for gaming, the architecture translates to other high-performance computing applications.

2.1   Overview

Over the past decade, microprocessor design has increasingly focused on multi-core architectures [9]. A multi-core architecture, when used

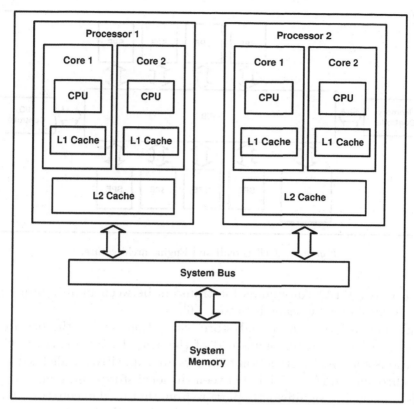

Figure 1. Example x86 dual-core, dual-processor system [14].

with multi-threaded applications, can keep pace with the performance demands of emerging consumer and scientific applications.

The multi-core architecture is the most popular conventional stored-program architecture (e.g., the x86 dual-core, dual-processor system in Figure 1). However, the architecture is limited by the von Neumann bottleneck – data transfer between processors and memory. Although processor speed has increased by more than two orders of magnitude over the last twenty years, application performance is still limited by memory latency rather than peak compute capability or peak bandwidth [6]. The industry response has been to move memory closer to the processor through on-board and local caches.

Similar to the conventional stored-program architecture, the CBE stores frequently-used code and data closer to the execution units. The implementation, however, is quite different [11]. Instead of the cache hierarchy implemented in traditional architectures, CBE uses direct mem-

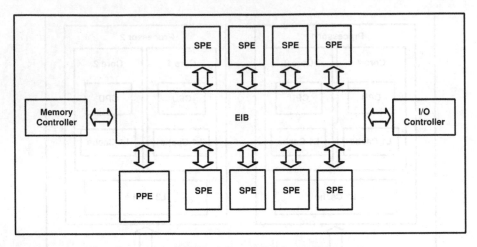

Figure 2. Cell Broadband Engine architecture.

ory access (DMA) commands for transfers between main storage and private local memory called "stores" [6].

Instruction fetches and load/store instructions access the private local store of each processor instead of the shared main memory. Computations and data/instruction transfers are explicitly parallelized using asynchronous DMA transfers between the local stores and main storage. This strategy is a significant departure from that used in conventional architectures that, at best, obtains minimal independent memory accesses. In the cell architecture, the DMA controller processes memory transfers asynchronously while the processor operates on previously-transferred data.

Another major difference is the use of single-instruction, multiple data (SIMD) computations. SIMD computations, first employed in large-scale supercomputers, support data-level parallelism [10]. The vector processors that implement SIMD perform mathematical operations on multiple data elements simultaneously. This differs from the majority of processor designs, which use scalar processors to process one data item at a time and rely on instruction pipelining and other techniques for speed-up. While scalar processors are primarily used for general purpose computing, SIMD-centered architectures are often used for data-intensive processing in scientific and multimedia applications [4].

2.2   Architectural Details

Figure 2 presents a block diagram of the CBE architecture. The CBE incorporates nine independent core processors: one PowerPC processor element (PPE) and eight synergistic processor elements (SPEs). The

PPE is a 64-bit PowerPC chip with 512 KB cache that serves as the controller for a cell [5]. The PPE runs the operating system and the top-level control thread of an application; most computing tasks are off-loaded to the SPEs [2].

Each SPE contains a processor that uses the DMA and SIMD schemes. An SPE can run individual applications or threads; depending on the PPE scheduling, it can work independently or collectively with other SPEs on computing tasks. Each SPE has a 256 KB local store, four single-precision floating point units and four integer units capable of operating at 4 GHz. According to IBM [2], given the right task, a single SPE can perform as well as a high-end desktop CPU.

The element interface bus (EIB) incorporates four 128-bit concentric rings for transferring data between the various units [5]. The EIB can support more than 100 simultaneous DMA requests between main memory and SPEs with an internal bandwidth of 96 bytes per cycle. The EIB was designed specifically for scalability – because data travels no more than the width of one SPE, additional SPEs can be incorporated without changing signal path lengths, thereby increasing only the maximum latency of the rings [15].

The memory controller interfaces the main storage to the EIB via two Rambus extreme data rate (XDR) memory channels that provide a maximum bandwidth of 25.6 GBps [2]. An I/O controller serves as the interface between external devices and the EIB. Two Rambus FlexIO external I/O channels are available that provide up to 76.8 GBps accumulated bandwidth [5]. One channel supports non-coherent links such as device interconnect; the other supports coherent transfers (or optionally an additional non-coherent link). The two channels enable seamless connections with additional cell systems.

The PS3 cell architecture is presented in Figure 3. Although it was designed for use in gaming systems, it offers massive computational power for scientific applications at low cost [2]. The architecture permits the interconnection of multiple systems for computationally-intensive problems. When additional computing power is required, additional systems are easily plugged in to expand parallel processing.

3.   Password Cracking

We conducted a series of experiments involving password generation (using MD5 hashing) to evaluate the feasibility of using the PS3 for password cracking. The goal was to evaluate the PS3 as a viable, cost-effective option compared with the Intel and AMD systems that are currently used for password cracking.

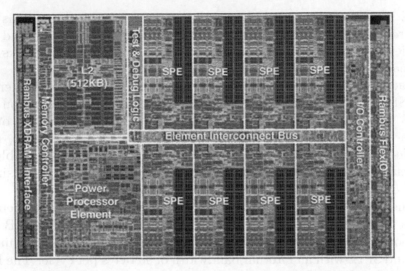

Figure 3. Die photograph of the PS3 cell architecture [2].

3.1   Experimental Setup

The experiments involved a PS3; an Intel Xeon 3.0 GHZ dual-core, dual-processor system with 2 GB RAM; and an AMD Phenom 2.5 GHz quad-core system with 4 GB RAM. All three systems ran the Fedora 9 Linux OS. MD5 code was written in C/C++ according to RFC 1321 [8] for use on the Intel and AMD systems. The algorithm was then ported to SIMD vector code for execution on the PS3. The CBE Software Development Kit [6] provided the necessary runtime libraries for compiling and executing code on the PS3.

Password generation involved two steps. The first step generated strings using a 72 character set (26 uppercase letters, 26 lowercase letters, 10 numbers and 10 special characters). The second step applied the MD5 cryptographic algorithm to create the passwords. The functionality of our MD5 code was verified by comparing generated passwords with the Linux MD5 generator. The password space was divided evenly among system processors and a strict brute-force implementation was used. Experiments were performed using password lengths of four, five, six and eight.

3.2   Experimental Results

Three metrics were used to evaluate password cracking performance: (i) passwords per second, (ii) computational time, and (iii) cost efficiency (i.e., "bang per buck"). The number of passwords per second was computed as the mean number of passwords generated during specific time

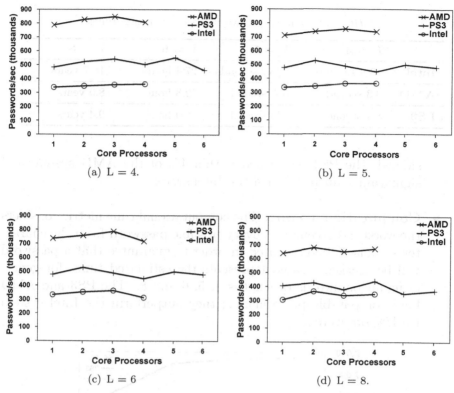

Figure 4. Number of passwords generated per second.

intervals (using a 95% confidence interval). The compute time was the estimated time required to generate the entire password space. The cost efficiency was calculated by dividing the number of passwords generated per second by the cost (in dollars) of the computing platform.

- **Passwords/Second:** The passwords per second metric was calculated for each independent processor. The Intel and AMD systems each use the equivalent of four processors while the PS3 uses six SPEs for computations (two SPEs are reserved for testing/overhead and are not easily accessible). Figure 4 presents the results obtained for various password lengths (L = 4, 5, 6 and 8). For the case where L = 4 (Figure 4(a)), the AMD system generated roughly 800K passwords/s for each processor; the PS3 generated approximately 500K passwords/s per processor and the Intel system 350K passwords/s per processor. As a system, AMD yielded close to 3.2M passwords/s, the PS3 generated 3.1M passwords/s, and Intel 1.4M passwords/s. For the four password lengths con-

Table 1.   Estimated password space generation times.

	L = 4	L = 5	L = 6	L = 8
Intel	18.8 seconds	1,377 seconds	27.4 hours	18.7 years
AMD	8.3 seconds	657 seconds	12.8 hours	8.9 years
PS3	8.4 seconds	649 seconds	12.9 hours	9.4 years

sidered, the PS3 performed within 4% of the AMD system and significantly outperformed the Intel system.

- **Computational Time:** The computational time metric considers the worst-case scenario – every possible password from the character set has to be generated in order to guarantee that a password will be cracked. Table 1 presents the estimated times for generating all passwords of lengths 4, 5, 6 and 8. The PS3 and AMD have comparable performance; they outperform the Intel system by 47% on average.

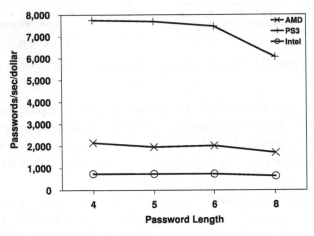

Figure 5.   Cost efficiency results.

- **Cost Efficiency:** The cost efficiency metric compares the "bang per buck" obtained for the three password cracking systems (Figure 5). For passwords of length four, the PS3 computes 7,700 passwords/s/$ compared with 2,100 passwords/s/$ for AMD and 750 passwords/s/$ for Intel. The results are based on 2009 prices of $400 for the PS3, $1,500 for AMD and $1,900 for Intel. This metric clearly demonstrates the cost efficiency of the PS3, which, on the average, outperforms the PS3 by a factor of 3.7 and Intel by a factor of 10.1.

3.3 Analysis of Results

The experimental results demonstrate that the PS3 is a viable option for password cracking. Techniques such as using separate SPEs for generating character strings and applying the MD5 algorithm, and implementing code optimization techniques (e.g., bit shifting operations) would certainly improve password cracking performance. Since the intent was not to demonstrate the full potential and computing power of the PS3, the SIMD vector code used in the experiments was not optimized. Significant computational gains can be achieved when the SIMD vector code is optimized for the PS3. Nevertheless, as the results demonstrate, the PS3 is very effective even without code optimization.

A 2007 study showed that a PS3 generates the key space for cryptographic algorithms at twenty times the speed of an Intel Core 2 Duo processor [3]. The ease of expanding computations across multiple PS3s make a multi-system framework an attractive option for large-scale password cracking efforts. Indeed, a dozen PS3 systems purchased for about $5,000 can provide the computing power of a cluster of conventional workstations costing in excess of $200,000.

4. Distributed Password Cracking Framework

Criminals are increasingly using encryption to hide evidence and other critical information. Most law enforcement agencies do not have the technical resources or the time to apply cryptanalytic techniques to recover vital evidence. The distributed framework presented in this section is intended to provide law enforcement agents in a large geographic region with high-performance computing resources to crack encrypted archives and applications.

The distributed password cracking framework shown in Figure 6 is designed for use by a major U.S. law enforcement agency. The design enables agents located in field offices to submit encrypted files to the Scheduler/Controller for password cracking. The Scheduler/Controller manages system overhead and assigns jobs to the PS3 Cluster based on job priority, evidence type and resource availability. The high-performance computing resources then go to work performing cryptanalysis on the submitted evidence.

Once a job is complete, details are recorded in the Repository and the submitting agent is notified about the results. The Repository serves as a vault for evidence and as a cross-referencing facility for evidence and

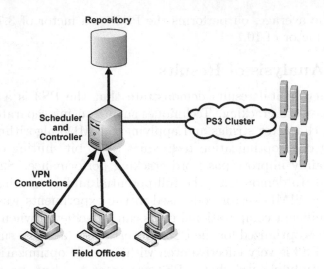

Figure 6.   Distributed password cracking framework.

case files. Over time, common passwords and trends will emerge that can help accelerate password cracking.

The framework incorporates special security requirements that satisfy law enforcement constraints. Network channels and communications are protected using encryption and VPN tunnels. Strict identification, authentication and authorization are implemented to control access to evidence. Additionally, hashing algorithms are incorporated to ensure the integrity of the processed evidence.

Figure 7 presents a detailed view of the framework. Law enforcement agents interact with the system using the Client Module. The application programming interface (API) of the Client Module is designed to support GUI applications built on C++, Java or PHP/HTML. Using secure connections, law enforcement agents may submit new evidence, check the status of submitted evidence, and query the system for case data and statistical information.

The Command and Control Module manages system access and facilitates user queries and evidence submission. It runs on a dedicated, robust server that maintains secure access to the password cracking cluster. A Cipher Module running on each PS3 reports workload and resource availability data to the Job Scheduling Module. The Job Scheduling Module submits a job to an appropriate Cipher Module for execution. The Cipher Module manages cryptanalysis by distributing the password cracking effort between the SPEs. The key space is divided over the "allocated" or available cycles for a particular encrypted file. This could

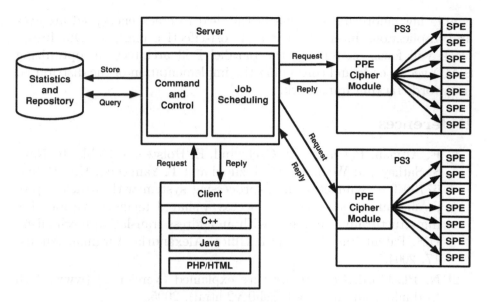

Figure 7. Framework components.

involve one SPE in one PS3, all the SPEs in one PS3, or multiple SPEs in multiple PS3s.

As an example, consider a Microsoft Excel 2003 spreadsheet that has been password protected. The law enforcement agent submits the file for processing. Based on the workload and the available cycles of the PS3s in the cluster, the Cipher Module allocates "chunks" of the key space for SPEs to start generating. As the keys are generated, they are encrypted using the appropriate algorithm (e.g., RC4 for Microsoft Excel 2003) and compared with the submitted evidence to locate a match. When the password is found, the results are returned to the law enforcement agent and the statistics and records are stored in the Repository.

To enable efficient password cracking, the Cipher Modules engage brute force methods augmented with dictionary word lists, previously encountered passwords and variations of commonly-used passwords. In addition, rainbow tables with pre-computed hash values and other cipher exploitation techniques may be leveraged [13]. The distributed framework alleviates the need for each field office to separately maintain expensive equipment for password cracking.

5. Conclusions

Law enforcement agencies can leverage the high-performance, extensibility and low cost of PS3 systems for password cracking in digital forensic investigations. The distributed PS3-based framework is designed to enable law enforcement agents in the field to crack encrypted archives and applications in an efficient and cost-effective manner. Our future work will focus on refining the parallel PS3 architecture, optimizing SIMD vector code and continuing the implementation of the distributed password cracking framework.

References

[1] E. Altman, P. Capek, M. Gschwind, H. Hofstee, J. Kahle, R. Nair, S. Sathaye, J. Wellman, M. Suzuoki and T. Yamazaki, U.S. Patent 6779049 – Symmetric multi-processing system with attached processing units being able to access a shared memory without being structurally configured with an address translation mechanism, U.S. Patent and Trademark Office, Alexandria, Virginia, August 17, 2004.

[2] N. Blachford, Cell architecture explained (Version 2) (www.blach ford.info/computer/Cell/Cell0_v2.html), 2005.

[3] N. Breese, Crackstation – Optimized cryptography on the PlayStation 3, Security-Assessment.com, Auckland, New Zealand (www.se curity-assessment.com/files/presentations/crackstation-njb-bheu08 -v2.pdf), 2007.

[4] M. Gschwind, The cell architecture, IBM Corporation, Armonk, New York (domino.research.ibm.com/comm/research.nsf/pages/r.a rch.innovation.html), 2007.

[5] IBM Corporation, Cell Broadband Engine Architecture (Version 1.02), Armonk, New York, 2007.

[6] IBM Corporation, IBM Software Development Kit for Multicore Acceleration (Version 3), Armonk, New York, 2007.

[7] E. Nakashima, In child porn case, a digital dilemma, *The Washington Post*, January 16, 2008.

[8] R. Rivest, RFC 1321: The MD5 Message-Digest Algorithm, Laboratory for Computer Science, Massachusetts Institute of Technology, Cambridge, Massachusetts (www.ietf.org/rfc/rfc1321.txt), 1992.

[9] A. Shimpi, Understanding the cell microprocessor, AnandTech, Minden, Nevada (www.anandtech.com/cpuchipsets/showdoc.aspx? i=2379), 2005.

[10] J. Stokes, SIMD architecture, Ars Technica (arstechnica.com/artic les/paedia/cpu/simd.ars), 2000.

[11] J. Stokes, Introducing the IBM/Sony/Toshiba cell processor Part I: The SIMD processing units, Ars Technica (arstechnica .com/articles/paedia/cpu/cell-1.ars), 2005.

[12] M. Suzuoki, T. Yamazaki, C. Johns, S. Asano, A. Kunimatsu and Y. Watanabe, U.S. Patent 6809734 – Resource dedication system and method for a computer architecture for broadband networks, U.S. Patent and Trademark Office, Alexandria, Virginia, October 26, 2004.

[13] C. Swenson, *Modern Cryptanalysis: Techniques for Advanced Code Breaking*, Wiley, Indianapolis, Indiana, 2008.

[14] T. Tian and C. Shih, Software techniques for shared-cache multi-core systems, Intel Corporation, Santa Clara, California (software.intel.com/en-us/articles/software-techniques-for-shared-cache-multi-core-systems), 2007.

[15] D. Wang, The cell microprocessor, Real World Technologies, Colton, California (www.realworldtech.com/page.cfm?ArticleID=RWT021005084318), 2005.

VI

INVESTIGATIVE TECHNIQUES

Chapter 17

A COST-EFFECTIVE MODEL FOR DIGITAL FORENSIC INVESTIGATIONS

Richard Overill, Michael Kwan, Kam-Pui Chow, Pierre Lai and Frank Law

Abstract   Because of the way computers operate, every discrete event potentially leaves a digital trace. These digital traces must be retrieved during a digital forensic investigation to prove or refute an alleged crime. Given resource constraints, it is not always feasible (or necessary) for law enforcement to retrieve all the related digital traces and to conduct comprehensive investigations. This paper attempts to address the issue by proposing a model for conducting swift, practical and cost-effective digital forensic investigations.

Keywords: Investigation model, Bayesian network

1.   Introduction

A digital forensic investigation involves the application of a series of processes on digital evidence such as identification, preservation, analysis and presentation. During the analysis process, digital forensic investigators reconstruct events in order to evaluate the truth of the forensic hypotheses related to the crime or incident based on the digital traces that have been identified and retrieved [2]. Due to inherent technological complexities, the identification and retrieval of digital traces cover a variety of techniques such as cryptography, data carving and data reconstruction. Each technique has a different level of complexity and, therefore, a different resource cost (e.g., expertise, time and tools).

Unlike physical events that are continuous, digital events are discrete and occur in temporal sequence [3]. Because of the discrete nature, it is possible to quantify the retrieval costs of individual digital traces. However, in the absence of a suitable model for digital forensic investigations, most investigators attempt to conduct a comprehensive retrieval

G. Peterson and S. Shenoi (Eds.): Advances in Digital Forensics V, IFIP AICT 306, pp. 231–240, 2009.

of all related digital traces despite the substantial costs associated with retrieving all the traces.

A more effective technique is to focus only on the digital traces that can be extracted in a cost-effective manner. The reasons are that investigating different digital traces requires resources (e.g., expertise, time and tools) in different amounts, and that the traces found have different evidentiary weights with respect to proving a hypothesis. The limited resources available for an investigation renders exhaustive search approaches impractical [4]. Consequently, digital forensic investigators who endeavor to retrieve all the traces – especially those that are not sufficient to prove the hypotheses – waste valuable resources.

This paper describes a model for conducting swift, practical and cost-effective digital forensic investigations. The model considers the retrieval costs of digital traces and incorporates a permutation analysis.

2.   Preliminaries

Using the collective experience and judgment of digital forensic investigators, it is possible to rank the relative costs of investigating each trace T_i $(i = 1 \ldots m)$. The relative costs may be estimated in terms of their resource requirements (person-hours, access to specialized equipment, etc.) using standard business accounting procedures. The relative costs can be ranked $T_1 \leq T_2 \leq \ldots \leq T_m$ without any loss of generality. As a direct consequence of this ranking, the minimum cost path for the overall investigation is uniquely identified.

Our focus is on digital traces residing on a hard disk. If the seized computer has sufficient storage, all the digital traces can be retrieved. If all the traces T_i $(i = 1 \ldots m)$ are retrieved, the minimum cost path is the permutation $[T_1 T_2 \ldots T_m]$. An example of a permutation path is shown in Figure 1.

The number of possible paths at each step is given by $m!$ This is a direct consequence of the fact that the problem of selecting the next available trace from an ordered permutation of m distinct traces is isomorphic to the problem of selecting the next object from a collection of m identical objects.

In order to save time and conserve resources, it is useful to determine early in an investigation whether or not the investigation should continue. This requires an estimation of the cumulative evidentiary weight associated with the investigation as $W = \sum_{i=1}^{m} w_i$, where the (scaled) relative fractional evidentiary weight w_i of each trace T_i is either assigned by an expert or, by default, is set to one. The weight assignment process has to be undertaken only once as a preprocessing step for each distinct

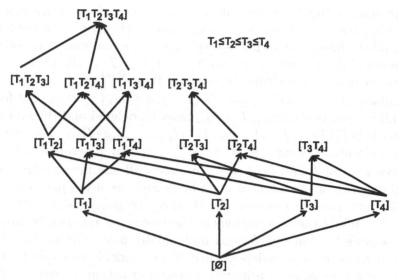

Figure 1. Path diagram with four traces.

digital crime template. If the cumulative estimate W is sufficiently close to one, the *prima facie* of the case can probably be established. Otherwise, it is unlikely that the available digital traces are sufficient to support the case.

The difference between W and one provides a "cut-off" condition that an investigator can use to avoid identifying all the traces exhaustively. The cut-off state is illustrated using the following example. Suppose that email exchanges between the culprit and victim are vital to an investigation. The forensic goal is to confirm that the computer, which was under the culprit's control, had been used to send and receive the emails in question. Assume that the evidentiary traces are T_1, T_2, T_3, T_4, T_5 with evidentiary weights 0.05, 0.10, 0.15, 0.2, 0.35, respectively; and the evidentiary threshold is 0.85. Therefore, if all the traces are retrieved, then the estimated total evidentiary weight is 0.85, which indicates a strong case. On the other hand, if trace T_1 is not found, the overall evidentiary weight is 0.8, which is a 6% falloff. If both T_1 and T_2 are missing, the overall weight becomes 0.7, an 18% falloff. At this point, the forensic investigator should consider suspending the investigation as the prospect of successful prosecution is slim.

3.   Missing Traces

Since a computer may not have sufficient storage, there is always the chance that some traces may be missing or overwritten. Thus, it may not be possible for an investigator to ascertain all the trace evidence pertaining to a case. Suppose a certain trace T_j $(1 \leq j \leq m)$ is not found. Then, all the investigative paths involving T_j are removed from the path diagram and the minimum cost path becomes $[T_1 T_2 \ldots T_{j-1} T_{j+1} \ldots T_m]$. The estimate of the evidentiary weight is $W = \sum_{i \neq j}^{m} w_i$.

Similarly, if two traces T_j and T_k $(1 \leq j; k \leq m; j \leq k)$ are not found, then all the paths involving T_j or T_k must be deleted and the minimum cost path is $[T_1 T_2 \ldots T_{j-1} T_{j+1} \ldots T_{k-1} T_{k+1} \ldots T_{m-1} T_m]$. The estimate of the evidentiary weight is $W = \sum_{i \neq j,k}^{m} w_i$. In general, if a total of k traces are not found $(1 \leq k < m)$, then all the investigative paths containing any of the k traces must be deleted from the path diagram.

It is important to consider the issue of the independence of digital traces T_i. While the observations of the traces are necessarily independent because they are performed individually *post mortem*, the digital traces must be created independently if the model is to retain its validity. Since it is possible in principle for one user action to create multiple digital traces T_i (which are not mutually independent), care must be taken to ensure the independence of the expected digital traces when selecting the set of traces.

4.   Investigation Model

The model for conducting cost-effective digital forensic investigations has two phases.

Phase 1 (Preprocessing – Detecting Traces)

- Enumerate the set of traces expected to be present based on the type of crime suspected.
- Assign relative investigation costs to each expected trace.
- Rank the expected traces in order of increasing relative investigation costs.
- Assign relative evidentiary weights w_i to each ranked trace.
- Rank the expected traces within each cost band in order of decreasing relative evidentiary weight.
- Set the cumulative evidentiary weight estimate W to zero.
- Set the total of the remaining available weights W_{rem} to one.
- For each expected trace taken in ranked order:
  - Search for the expected trace.
  - Subtract the relative evidentiary weight w_i of the trace from W_{rem}.

- If the expected trace is retrieved, add its relative evidentiary weight w_i to W.

- If W is sufficiently close to one, proceed to Phase 2.

- If $W + W_{rem}$ is not close enough to one, abandon the forensic investigation.

Phase 2 (Bayesian Network – Analyzing Traces)

- Run and analyze the Bayesian network model for the crime hypothesis using the retrieved traces as evidence (as described in [6]).

5.   BitTorrent Case Study

This section uses a BitTorrent case study [6] to demonstrate the cost-effective digital forensic investigation model.

Figure 2 shows a Bayesian network with eighteen expected evidence traces (E_i) and their relationships to the five hypotheses (H_i). The Bayesian network is constructed by enumerating every path through which an evidentiary trace could have been produced and assigning it a probability.

The ideal case, in which all eighteen evidence traces are retrieved, is shown in Table 1. Note that each piece of trace evidence E_i is ranked according to its cost T_j.

The actual case, corresponding to a situation where two of the expected traces (E_8 and E_{14}) are missing, is shown in Table 2.

A potential complication involving the proposed investigation model should be noted. A trace could initially be assigned a low cost; however, upon further consideration, it could be determined that the cost is much higher. Examples of such a situation are a file that turns out to be protected by encryption or a partition that turns out to be deleted. In such cases, the cost of investigating the trace must be revised and all the traces must be ranked again based on the revised costs. This is necessary to maintain the minimum cost strategy for the investigation.

Constructing a Bayesian network model corresponding to an investigation requires the definition of the overall structure of the network, including the hierarchy of hypotheses and the associated posterior digital evidence (or traces) whose presence or absence determines the prior probabilities of the corresponding hypotheses. Next, numerical values are assigned to the prior probabilities. Traditionally, forensic investigators assign the prior probabilities based their expertise and experience.

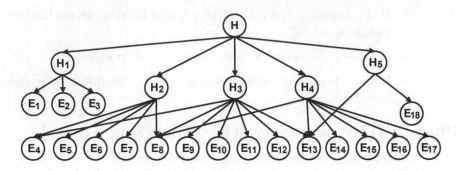

HYPOTHESES:

H – The seized computer was used as the initial seeder to share the pirated file on a
      BitTorrent network

H₁ – The pirated file was copied from the seized optical disk to the seized computer

H₂ – A torrent file was created from the copied file

H₃ – The torrent file was sent to newsgroups for publishing

H₄ – The torrent file was activated and the computer connected to the tracker server

H₅ – The connection between the seized computer and the tracker was maintained

EVIDENCE:

E₁ – Modification time of the destination file is the same as that of the source file

E₂ – Creation time of the destination file is after its modification time

E₃ – Hash value of the destination file matches that of the source file

E₄ – BitTorrent client software is installed on the computer

E₅ – File link for the shared file is found

E₆ – Shared file exists on the hard disk

E₇ – Torrent file creation record is found

E₈ – Torrent file exists on the hard disk

E₉ – Peer connection information is found

E₁₀ – Tracker server login record is found

E₁₁ – Torrent file activation time is corroborated by its MAC time and link file

E₁₂ – Internet history record of the publishing website is found

E₁₃ – Internet connection is available

E₁₄ – Cookie of the publishing website is found

E₁₅ – URL of the publishing website is stored in the web browser

E₁₆ – Web browser software is found

E₁₇ – Internet cache record of the publishing of the torrent file is found

E₁₈ – Internet history record of the tracker server connection is found

Figure 2.   Bayesian network model for the BitTorrent case.

However, these assessments have been challenged in judicial proceedings
primarily on the grounds that they are non-rigorous and subjective.

These challenges can be countered if a rigorous analytic procedure
is used to quantitatively assign the prior probabilities. A promising

Table 1. Traces, relative costs and weights for the ideal BitTorrent case.

Trace		Rel. Cost	Rel. Wt.	W	W_{rem}
	Initial Values			0	1
T_1 (E_6)	Shared file exists on the hard disk	1	2/18	2/18	16/18
T_2 (E_1)	Modification time of the destination file is the same as that of the source file	1	1/18	3/18	15/18
T_3 (E_2)	Creation time of the destination file is after its modification time	1	1/18	4/18	14/18
T_4 (E_3)	Hash value of the destination file matches that of the source file	1	1/18	5/18	13/18
T_5 (E_8)	Torrent file exists on the hard disk	1	1/18	6/18	12/18
T_6 (E_{16})	Web browser software is found	1	1/18	7/18	11/18
T_7 (E_5)	File link for the shared file is found	1	0.5/18	7.5/18	10.5/18
T_8 (E_{15})	URL of the publishing website is stored in the web browser	1	0.5/18	8/18	10/18
T_9 (E_7)	Torrent file creation record is found	1.5	2/18	10/18	8/18
T_{10} (E_{13})	Internet connection is available	1.5	2/18	12/18	6/18
T_{11} (E_{10})	Tracker server login record is found	1.5	0.5/18	12.5/18	5.5/18
T_{12} (E_{12})	Internet history record of the publishing website is found	1.5	0.5/18	13/18	5/18
T_{13} (E_{14})	Cookie of the publishing website is found	1.5	0.5/18	13.5/18	4.5/18
T_{14} (E_{17})	Internet cache record of the publishing of the torrent file is found	1.5	0.5/18	14/18	4/18
T_{15} (E_{18})	Internet history record of the tracker server connection is found	1.5	0.5/18	14.5/18	3.5/18
T_{16} (E_4)	BitTorrent client software is installed on the computer	2	2/18	16.5/18	1.5/18
T_{17} (E_{11})	Torrent file activation time is corroborated by its MAC time and link file	2	1/18	17.5/18	0.5/18
T_{18} (E_9)	Peer connection information is found	2	0.5/18	1	0

Table 2.   Traces, relative costs and weights for the actual BitTorrent case.

	Trace	Rel. Cost	Rel. Wt.	W	W_{rem}
	Initial Values			**0**	**1**
T_1 (E_6)	Shared file exists on the hard disk	1	2/18	2/18	16/18
T_2 (E_1)	Modification time of the destination file is the same as that of the source file	1	1/18	3/18	15/18
T_3 (E_2)	Creation time of the destination file is after its modification time	1	1/18	4/18	14/18
T_4 (E_3)	Hash value of the destination file matches that of the source file	1	1/18	5/18	13/18
T_5 (E_8)	Torrent file exists on the hard disk *(missing)*	1	1/18	5/18	12/18
T_6 (E_{16})	Web browser software is found	1	1/18	6/18	11/18
T_7 (E_5)	File link for the shared file is found	1	0.5/18	6.5/18	10.5/18
T_8 (E_{15})	URL of the publishing website is stored in the web browser	1	0.5/18	7/18	10/18
T_9 (E_7)	Torrent file creation record is found	1.5	2/18	9/18	8/18
T_{10} (E_{13})	Internet connection is available	1.5	2/18	11/18	6/18
T_{11} (E_{10})	Tracker server login record is found	1.5	0.5/18	11.5/18	5.5/18
T_{12} (E_{12})	Internet history record of the publishing website is found	1.5	0.5/18	12/18	5/18
T_{13} (E_{14})	Cookie of the publishing website is found *(missing)*	1.5	0.5/18	12/18	4.5/18
T_{14} (E_{17})	Internet cache record of the publishing of the torrent file is found	1.5	0.5/18	12.5/18	4/18
T_{15} (E_{18})	Internet history record of the tracker server connection is found	1.5	0.5/18	13/18	3.5/18
T_{16} (E_4)	BitTorrent client software is installed on the computer	2	2/18	15/18	1.5/18
T_{17} (E_{11})	Torrent file activation time is corroborated by its MAC time and link file	2	1/18	16/18	0.5/18
T_{18} (E_9)	Peer connection information is found	2	0.5/18	16.5/18	0

approach is to use complexity theory [7]. Essentially, every path by which an evidential trace could have been produced is enumerated, and the probability associated with each path is evaluated using techniques from complexity theory.

We illustrate the approach using an example from the BitTorrent case. In particular, we evaluate the prior probability that hypothesis H_2 is true given that trace evidence E_8 (i.e., T_5) is found (Figure 2). E_8 is the evidence that the torrent file is present on the hard disk of the seized computer.

Three scenarios that result in the presence of the torrent file are:

- The file was placed on the seized computer by a covert malware process.

- The file was copied or downloaded to the seized computer from some other source.

- The file was created on the seized computer from the pirated file.

Assume that a state-of-the-art, anti-malware scan reveals the presence of a Trojan with a probability of approximately 0.98 [5]. Additionally, a thorough, careful inventory of the local networked drives and portable storage media reveals the presence of a source copy of the torrent file with a probability greater than 0.95. Furthermore, a high-quality search engine detects the presence of a downloadable copy of the torrent file with a similar probability [1]. As a result, the probability that the torrent file was created *in situ* on the hard disk of the seized computer is at least 0.88. The error bars for the assigned probabilities are derived assuming that the errors are normally distributed. Based on these assignments, we obtain a probability value of 0.94 ± 0.06.

6.   Conclusions

The proposed two-phase digital forensic investigation model achieves the twin goals of reliability and cost-effectiveness by incorporating a pre-processing phase, which runs in parallel with the data collection phase. The evidentiary weighting and cost ranking of the expected traces, which are undertaken only once for all similar investigations, enable the lowest cost traces to be examined first. This means that the "best-case" and "worst-case" scenarios can be processed efficiently. The combined use of evidentiary weights and ranked costs enables an ultimately futile investigation to be detected early and abandoned using only low cost traces. By the same token, an investigation that would ultimately prove to be unsuccessful could be halted before any high cost traces are investigated.

This model performs best in cases where the distributions of evidentiary weights versus costs are skewed towards low costs, and it performs the worst when the distributions are skewed towards high costs. In the average case, where the distribution is essentially unskewed (or even uniform), the model exhibits intermediate performance. However, it should be noted that, even in the most pathological cases, the performance would not be significantly worse than the current exhaustive or random search for traces.

One of the main advantages of the model is that it offers the possibility of creating templates of expected traces and their associated costs and evidentiary weights for every type of digital crime. These templates can provide investigators with benchmarks for calibrating their investigative procedures, and also offer novice investigators with an investigative model that can be adopted in its entirety.

References

[1] S. Brin and L. Page, The anatomy of a large-scale hypertextual web search engine, *Computer Networks and ISDN Systems*, vol. 30(1-7), pp. 107–117, 1998.

[2] B. Carrier and E. Spafford, Defining event reconstruction of digital crime scenes, *Journal of Forensic Sciences*, vol. 49(6), pp. 1291–1298, 2004.

[3] E. Casey, *Digital Evidence and Computer Crime: Forensic Science, Computers and the Internet*, Academic Press, London, United Kingdom, 2004.

[4] Joint Committee on Human Rights, Counter-Terrorism Policy and Human Rights: Terrorism Bill and Related Matters, Third Report of Session 2005-06, HL Paper 75-I, HC 561-I, House of Lords, House of Commons, London, United Kingdom, 2005.

[5] Kaspersky Lab, Free online virus scanner, Woburn, Massachusetts (www.kaspersky.com/virusscanner).

[6] M. Kwan, K. Chow, F. Law and P. Lai, Reasoning about evidence using Bayesian networks, in *Advances in Digital Forensics IV*, I. Ray and S. Shenoi (Eds.), Springer, Boston, Massachusetts, pp. 275–289, 2008.

[7] S. Lloyd, Measures of complexity: A non-exhaustive list, *IEEE Control Systems*, vol. 21(4), pp. 7–8, 2001.

Chapter 18

ANALYSIS OF THE DIGITAL EVIDENCE PRESENTED IN THE YAHOO! CASE

Michael Kwan, Kam-Pui Chow, Pierre Lai, Frank Law and Hayson Tse

Abstract   The "Yahoo! Case" led to considerable debate about whether or not an IP address is personal data as defined by the Personal Data (Privacy) Ordinance (Chapter 486) of the Laws of Hong Kong. This paper discusses the digital evidence presented in the Yahoo! Case and evaluates the impact of the IP address on the verdict in the case. A Bayesian network is used to quantify the evidentiary strengths of hypotheses in the case and to reason about the evidence. The results demonstrate that the evidence about the IP address was significant to obtaining a conviction in the case.

Keywords: Yahoo! Case, digital evidence, Bayesian network, reasoning

1.   Introduction

Scientific conclusions based on evidence have been used for many years in forensic investigations. In making their assessments, investigators consider the available facts and the likelihood that they support or refute hypotheses related to a case. Investigators recognize that there is never absolute certainty and seek a degree of confidence with which to establish their hypotheses [2].

A forensic investigation determines the likelihood of a crime through the analysis and interpretation of evidence. To this end, a forensic investigation focuses on the validation of hypotheses based on the evidence and the evaluation of the likelihood that the hypotheses support legal arguments [6, 10, 12, 14, 15]. The likelihood represents the degree of belief in the truth of the associated hypothesis. It is typically expressed as a probability and probabilistic methods may be used to deduce the likelihood of a hypothesis based on the available evidence [7, 9].

G. Peterson and S. Shenoi (Eds.): Advances in Digital Forensics V, IFIP AICT 306, pp. 241–252, 2009.

A crime and its associated digital evidence are usually linked by sub-hypotheses. This paper uses a Bayesian network [8] to analyze and reason about the evidence in the well-known Yahoo! Case [3].

In the Yahoo! Case, Yahoo! Holdings (Hong Kong) Limited (Yahoo! HHKL) supplied IP address information to Chinese authorities that led to the conviction of Shi Tao, a Chinese journalist, for sending confidential state information to foreign entities. Shi Tao received a 10-year sentence for his crime.

Shi Tao's authorized representative in Hong Kong subsequently lodged a complaint with the Office of the Privacy Commissioner for Personal Data. The complaint maintained that Yahoo! HHKL disclosed Shi Tao's "personal data" to Chinese authorities, which was a breach of the Hong Kong Personal Data (Privacy) Ordinance.

The investigation by the Privacy Commissioner concluded that an IP address, on its own, does not constitute personal data [13]. The conclusion was based on the position that an IP address is unique to a specific computer not a person and, therefore, does not meet the definition of personal data. The Privacy Commissioner also held that no safe conclusion could be drawn that user data corresponding to the IP address belonged to a living individual as opposed to a corporate or unincorporated body, or that it was related to a real as opposed to a fictitious individual.

We use Bayesian network inference to assess the evidentiary weight of the IP address in the Yahoo! Case. Four scenarios are evaluated:

- Yahoo! HHKL and the ISP participate in the investigation; all the digital evidence is available.

- Yahoo! HHKL participates in the investigation; digital evidence regarding the IP address is received from Yahoo! HHKL. However, the ISP does not participate in the investigation.

- Yahoo! HHKL does not participate in the investigation; digital evidence regarding the IP address is not received from Yahoo! HHKL. However, the ISP participates in the investigation.

- Yahoo! HHKL and the ISP do not participate in the investigation; no digital evidence regarding the IP address is available.

Although an IP address, by itself, is not viewed as personal data, our analysis shows that it carried significant evidentiary weight in the Yahoo! Case. Our analysis is based on the "Reasons for Conviction" [4], and the Administrative Appeals Board decision [1] regarding the Report of the Hong Kong Privacy Commissioner published under Section 48(2) of the Personal Data (Privacy) Ordinance (Chapter 486) [13].

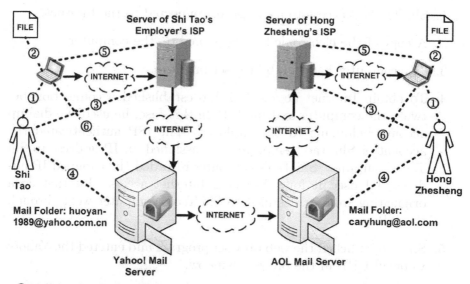

① Shi Tao (Hong Zhesheng) controls the computer.
② The attached file exists on Shi Tao's (Hong Zhesheng's) computer.
③ Shi Tao's employer's (Hong Zhesheng's) ISP subscription record.
④ Shi Tao's (Hong Zhesheng's) Yahoo! (AOL) email account registration record.
⑤ The computer connects to the ISP.
⑥ The web browser program displays the Yahoo! (AOL) email web page.

Figure 1. Entities and events in the Yahoo! Case.

2. Digital Evidence in the Yahoo! Case

In the Yahoo! Case, the Changsha Intermediate People's Court of Hunan Province convicted Shi Tao of providing state secrets to foreign entities. Based on the data provided by Yahoo! HHKL, the court determined that at approximately 11:32 pm on April 20, 2004, Shi Tao used a computer in his employer's office to access his personal email account (huoyan-1989@yahoo.com.cn) via the Yahoo! webmail interface and send some notes regarding a summary of a top-secret document issued by the Chinese Government to the email account of Hong Zhesheng (caryhung@aol.com) [13]. Shi Tao asked Hong Zhesheng, who resided in New York, to find a way to distribute the notes as quickly as possible without using Shi Tao's name [5].

Figure 1 shows the entities and events involved in the email transmission from Shi Tao to Hong Zhesheng. Based on this description, a digital forensic investigator would be required to ascertain the following facts:

1. Shi Tao had access to a computer connected to the Internet.

2. A copy of the electronic file was stored on the computer.

3. The computer had a web browser program.

4. To obtain Internet access, Shi Tao established a connection between the computer and the ISP. In this case, he used the dial-up account belonging to his employer. The ISP authenticated the account of Shi Tao's employer and assigned an IP address to Shi Tao's computer. Shi Tao's computer recorded the assigned IP address and used it for subsequent Internet access. Internet data originating from or destined to Shi Tao's computer went through the ISP.

5. Shi Tao launched the web browser program and entered the Yahoo! webmail URL in the browser window.

6. The web browser program sent an HTTP request to the Yahoo! mail server. When the requested web page was retrieved, it was displayed by the web browser program.

7. Shi Tao entered his user name and password to log into his email account. Based on the email subscription details, the Yahoo! mail server authenticated Shi Tao and allowed him to log into his email folder.

8. Shi Tao composed the email, attached the file and entered Hong Zhesheng's AOL email address. He then clicked the "Send" button to transmit the email along with the file attachment. Since Shi Tao used a web browser program to create the email, the email content was (possibly) cached in Shi Tao's computer.

9. The Yahoo! email server stored the email and the attachment, and placed it in the message queue for transmission to Hong Zhesheng's AOL email server via SMTP.

3.   Evaluation of Digital Evidence

In general, an investigation must clarify a number of issues before a case can be brought to court. These issues include whether or not a crime was committed, how the crime was committed, who committed the crime and whether or not there is a reasonable chance of conviction.

We use a Bayesian network to quantify the evidentiary strengths of hypotheses and to reason about evidence. A Bayesian network is a directed acyclic graph whose edges indicate dependencies between nodes.

Each node is accompanied by a conditional probability table (CPT) that describes the dependencies between nodes. In our work, the nodes correspond to hypotheses and the digital evidence associated with hypotheses. The edges connect each hypothesis to the evidence that should be present if the hypothesis is valid.

4. Bayesian Network

The first step in constructing a Bayesian network for analyzing digital evidence in the Yahoo! Case involves the definition of the primary hypothesis (H), the main issue to be determined. In the Yahoo! Case, the primary hypothesis is: "The seized computer was used to send the material document as an email attachment using a Yahoo! webmail account."

The next step is to define the possible states of the hypothesis (Yes, No and Uncertain). Probability values are then assigned to each state. Each of these values represents the prior probability that the hypothesis is in the specific state. The prior probability of H, $P(H)$, is assumed to be equal to (0.333, 0.333, 0.333), i.e., all three states are equally likely.

The hypothesis H is the root node in the Bayesian network. Sub-hypotheses that are causally dependent on the hypothesis assist in proving the hypothesis. The sub-hypotheses and the associated evidence and events are represented as child nodes in the Bayesian network.

Figure 1 lists six sub-hypotheses that support the primary hypothesis H in the Yahoo! Case. The six sub-hypotheses are:

- H_1: Linkage between the material document and the suspect's computer (Table 1).

- H_2: Linkage between the suspect and the computer (Table 2).

- H_3: Linkage between the suspect and the ISP (Table 3).

- H_4: Linkage between the suspect and the Yahoo! email account (Table 4).

- H_5: Linkage between the computer and the ISP (Table 5).

- H_6: Linkage between the computer and the Yahoo! email account (Table 6).

The evidence and events for the six sub-hypotheses are listed in Tables 1–6.

The states of the various sub-hypotheses are dependent on the state of H. Each sub-hypothesis, which is a child node of H, has an associated conditional probability table (CPT). The CPT contains the prior

Table 1. H_1: Linkage between the material document and the suspect's computer.

ID	Evidence Description	Type
DE1	Subject file exists on the computer	Digital
DE2	Last access time of the subject file is after the IP address assignment time by the ISP	Digital
DE3	Last access time of the subject file matches or is close to the sent time of the Yahoo! email	Digital

Table 2. H_2: Linkage between the suspect and the computer.

ID	Evidence Description	Type
PE1	Suspect was in physical possession of the computer	Physical
DE4	Files on the computer reveal the identity of the suspect	Digital

Table 3. H_3: Linkage between the suspect and the ISP.

ID	Evidence Description	Type
DE5	ISP subscription details match the suspect's particulars	Digital

Table 4. H_4: Linkage between the suspect and the Yahoo! email account.

ID	Evidence Description	Type
DE6	Subscription details of the Yahoo! email account match the suspect's particulars	Digital

Table 5. H_5: Linkage between the computer and the ISP.

ID	Evidence Description	Type
DE7	Configuration settings for the ISP Internet account are found on the computer	Digital
DE8	Log data confirms that the computer was powered up at the time the email was sent	Digital
DE9	Web program or email user agent program was found to be activated at the time the email was sent	Digital
DE10	Log data reveals the assigned IP address and the assignment time by the ISP to the computer	Digital
DE11	Assignment of the IP address to the suspect's account is confirmed by the ISP	Digital

Table 6. H_6: Linkage between the computer and the Yahoo! email account.

ID	Evidence Description	Type
DE12	Internet history logs reveal that the Yahoo! email account was accessed by the computer	Digital
DE13	Internet cache files reveal that the subject file was sent as an attachment via the Yahoo! email account	Digital
DE14	Yahoo! confirms the IP address of the Yahoo! email with the attached document	Digital

probabilities of the sub-hypothesis based on the state of the hypothesis. The probability values are typically assigned by digital forensic experts based on their subjective beliefs.

Table 7. Conditional probabilities of $H_1 \ldots H_6$.

H	$H_1 \ldots H_6$		
	Yes	No	Uncertain
Yes	0.60	0.35	0.05
No	0.35	0.60	0.05
Uncertain	0.05	0.05	0.90

We assume that all the sub-hypotheses ($H_1 \ldots H_6$) have the CPT values shown in Table 7. For example, an initial value of 0.6 is assigned for the situation where H and H_1 are Yes. This means that when the seized computer was used to send the material document as an email attachment using a Yahoo! webmail account, the probability that a linkage existed between the material document and the seized computer is 0.6. Additionally, there may be instances where it is not possible to confirm a Yes or No state for H_1 from the evidence although the seized computer was used to send the document. This uncertainty is modeled by assigning a probability of 0.05 to the Uncertain state.

After assigning conditional probabilities to the sub-hypotheses, the observable evidence and events related to each sub-hypothesis are added to the Bayesian network. For reasons of space, we only discuss Hypothesis H_1 (Linkage between the material document and the seized computer) in detail to demonstrate the use of a Bayesian network.

The evidence for H_1 that establishes the linkage between the material document and the seized computer includes: (i) the subject file exists on the computer; (ii) the last access time of the subject file is after the IP address assignment time by the ISP; and (iii) the last access time of

Table 8. Conditional probabilities of E_1, E_2, E_3.

H_1	E_1			E_2			E_3		
	Y	N	U	Y	N	U	Y	N	U
Y	0.85	0.15	0	0.85	0.15	0	0.85	0.12	0.03
N	0.15	0.85	0	0.15	0.85	0	0.12	0.85	0.03
U	0	0	1	0	0	1	0.03	0.03	0.94

the subject file matches or is close to the sent time of the Yahoo! email. Each node has the states: Yes (Y), No (N) and Uncertain (U).

The next step is to assign conditional probability values to the evidence. Table 8 shows the conditional probability values of evidence E_1, E_2 and E_3 given specific states of Hypothesis H_1.

After conditional probabilities are assigned to the entailing evidence, it is possible to propagate probabilities within the Bayesian network. In particular, the likelihood of H_1 is computed based on the observed probability values of evidence E_1, E_2 and E_3. The well-known MSBNX program [11] was used to propagate probabilities in the Bayesian network developed for the Yahoo! Case.

If evidence E_1, E_2 and E_3 have Yes states, then the digital forensic investigator can confirm that there is a likelihood of 99.6% that Hypothesis H_1 (Linkage between the material document and the suspect's computer) is true. Furthermore, based on the 99.6% likelihood for H_1, the investigator can also conclude that there is a 59.9% likelihood that H (The seized computer was used to send the material document as an email attachment using a Yahoo! webmail account) is true. Figure 2 shows the Bayesian network when E_1, E_2 and E_3 all have Yes states.

The same methodology is used to compute the likelihoods of the other five sub-hypotheses based on the probability values of the associated evidentiary nodes. Finally, the likelihoods of the six sub-hypotheses are used to compute the overall likelihood of the primary hypothesis.

5.   Impact of the IP Address

In order to assess the evidentiary weight of the IP address in the Yahoo! Case, we identify four scenarios that involve differing amounts of evidence provided to the Chinese authorities by Yahoo! HHKL and the ISP.

- **Scenario 1:** In this scenario, Yahoo! HHKL and the ISP participate in the investigation. When all the evidence (DE1–DE14

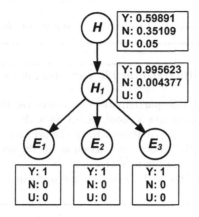

Figure 2. Probability distributions with E_1, E_2, E_3 = Yes.

and PE1) in Tables 1–6 is available and is true, the likelihood of Hypothesis H is 90.5%.

- **Scenario 2:** In this scenario, the ISP does not participate in the investigation. The evidentiary items DE5 (Table 3) and DE11 (Table 5) are missing. The corresponding likelihood of Hypothesis H is 88.1%.

- **Scenario 3:** In this scenario, Yahoo! HHKL does not participate in the investigation. The evidentiary items DE6 (Table 4) and DE14 (Table 6) are missing. The corresponding likelihood of Hypothesis H is 83.0%.

- **Scenario 4:** In this scenario, Yahoo! HHKL and the ISP do not participate in the investigation. Evidentiary items DE5 (Table 3), DE6 (Table 4), DE11 (Table 5) and DE14 (DE14) are missing. The corresponding likelihood of Hypothesis H is 78.7%.

Table 9 lists the four scenarios and their likelihoods. Note that the availability of the IP address affects the likelihood by 11.7%. In particular, the likelihood is 90.5% (very likely) when all the evidence is available, but it drops to 78.7% (probable) when evidence related to the IP address is not available. Although the IP address by itself does not reveal the identity of a specific user, it provides additional information that can further confirm the identity of the user.

The Reasons for Verdict [5] in the Yahoo! Case identified six primary facts:

- **Fact 1:** Shi Tao attended the press briefing and obtained the information.

Table 9.   Likelihood of Hypothesis *H*.

Scenario	Likelihood
Scenario 1: Yahoo! HHKL and the ISP participate in the investigation	90.5%
Scenario 2: Yahoo! HHKL participates in the investigation and confirms the IP address of the Yahoo! email with the attached document; the ISP does not participate in the investigation	88.1%
Scenario 3: Yahoo! HHKL does not participate in the investigation; the ISP participates in the investigation	83.0%
Scenario 4: Yahoo! HHKL and the ISP do not participate in the investigation	78.7%

- **Fact 2:** Shi Tao was present in the office of his employer at the material time.

- **Fact 3:** Shi Tao was the only employee who knew the information.

- **Fact 4:** The office of the employer was the registration address for the IP address.

- **Fact 5:** The IP address was assigned to the employer at the time the email was sent.

- **Fact 6:** The email was sent from the material IP address.

We developed a Bayesian network modeling these facts to evaluate the hypothesis: "Shi Tao sent the material email at the material time from the office of his employer." Experiments with the Bayesian network indicate that when all six facts are completely supported, the likelihood of Hypothesis *H* is 99.9%. However, when the IP address is missing (i.e., Facts 4–6 relating to the IP address are Uncertain), the overall likelihood drops to 14.9%, a reduction of 85.0%. This drop underscores the importance of the IP address in obtaining a conviction in the Yahoo! Case.

6.   Conclusions

Bayesian networks provide a powerful mechanism for quantifying the evidentiary strengths of investigative hypotheses and reasoning about evidence. The application of a Bayesian network to analyze digital evidence related to the Yahoo! Case demonstrates that the IP address was significant to obtaining a conviction. Investigators and prosecutors can use this technique very effectively to evaluate the impact of specific evidentiary items before a case is brought to court.

References

[1] Administrative Appeals Board, Shi Tao v. The Privacy Commissioner for Personal Data, Administrative Appeal No. 16 of 2007, Hong Kong (www.pcpd.org.hk/english/publications/files/Appeal _Yahoo.pdf), 2007.

[2] C. Aitken and F. Taroni, *Statistics and the Evaluation of Evidence for Forensic Scientists*, John Wiley and Sons, New York, 2004.

[3] R. Bascuas, Property and probable cause: The Fourth Amendment's principled protection of privacy, *Rutgers Law Review*, vol. 60(3), pp. 575–645, 2008.

[4] Changsha Intermediate People's Court of Hunan Province, Criminal Verdict, First Trial Case No. 29, Changsha Intermediate Criminal Division One Court, Changsha, China (www.globalvoicesonline .org/wp-content/ShiTao_verdict.pdf), 2005.

[5] Changsha Intermediate People's Court of Hunan Province, Reasons for Verdict, First Trial Case No. 29, Changsha Intermediate Criminal Division One Court, Changsha, China (www.pcpd.org.hk /english/publications/files/Yahoo_annex.pdf), 2005.

[6] R. Cook, I. Evett, G. Jackson, P. Jones and J. Lambert, A model for case assessment and interpretation, *Science and Justice*, vol. 38, pp. 151–156, 1998.

[7] P. Dawid, Statistics and the Law, Research Report No. 224, Department of Statistical Science, University College London, London, United Kingdom, 2004.

[8] F. Jensen, *An Introduction to Bayesian Networks*, Springer-Verlag, New York, 1996.

[9] M. Kwan, K. Chow, F. Law and P. Lai, Reasoning about evidence using Bayesian networks, in *Advances in Digital Forensics IV*, I. Ray and S. Shenoi (Eds.), Springer, Boston, Massachusetts, pp. 275–289, 2008.

[10] R. Loui, J. Norman, J. Altepeter, D. Pinkard, D. Craven, J. Linsday and M. Foltz, Progress on Room 5: A testbed for public interactive semi-formal legal argumentation, *Proceedings of the Sixth International Conference on Artificial Intelligence and Law*, pp. 207–214, 1997.

[11] Microsoft Research, MSBNx: Bayesian Network Editor and Tool Kit, Microsoft Corporation, Redmond, Washington (research.micro soft.com/adapt/MSBNx).

[12] J. Mortera, A. Dawid and S. Lauritzen, Probabilistic expert systems for DNA mixture profiling, *Theoretical Population Biology*, vol. 63(3), pp. 191–206, 2003.

[13] Office of the Privacy Commissioner for Personal Data, Report Published under Section 48(2) of the Personal Data (Privacy) Ordinance (Chapter 486), Report No. R07-3619, Hong Kong (www.pcpd.org .hk/english/publications/files/Yahoo_e.pdf), 2007.

[14] H. Prakken, C. Reed and D. Walton, Argumentation schemes and generalizations in reasoning about evidence, *Proceedings of the Ninth International Conference on Artificial Intelligence and Law*, pp. 32–41, 2003.

[15] D. Walton, Argumentation and theory of evidence, in *New Trends in Criminal Investigation and Evidence – Volume II*, C. Breur, M. Kommer, J. Nijboer and J. Reijntjes (Eds.), Intersentia, Antwerp, Belgium, pp. 711–732, 2000.

VII

LEGAL ISSUES

Chapter 19

ASSESSING THE LEGAL RISKS IN NETWORK FORENSIC PROBING

Michael Losavio, Olfa Nasraoui, Vincent Thacker, Jeff Marean, Nick Miles, Roman Yampolskiy and Ibrahim Imam

Abstract   This paper presents a framework for identifying the legal risks associated with performing network forensics on public networks. The framework is discussed in the context of the Gnutella P2P network protocol for which the legal issues related to authorized access have not yet been addressed.

Keywords: Network forensics, legal issues, authorized access, Gnutella protocol

1.   Introduction

The analysis of legal issues related to investigations of network misuse can help avoid misunderstandings about the application of law to computer and network conduct. An understanding of the issues is important for researchers who wish to avoid legal entanglements in the course of conducting their research and for investigators who are working on cases without the luxury of legal counsel.

An understanding of the application of law in this domain can also influence the course of litigation. For example, Craig Neidorf was prosecuted for the interstate transportation of stolen property (18 U.S.C. §2314) for the electronic BBS transmission and posting of a BellSouth 911 manual sent to him by a friend [6]. However, the prosecution was derailed when the government could not establish that the electronic version of the publicly available document was property of any type. One commentator attributed this to the conflict between traditional property law concepts and online activity. Similarly, Robert Morris Jr., the developer of the Internet worm, argued that he did not intend to cause damage with the release of his worm and was, therefore, not guilty un-

G. Peterson and S. Shenoi (Eds.): Advances in Digital Forensics V, IFIP AICT 306, pp. 255–266, 2009.
© IFIP International Federation for Information Processing 2009

Table 1.   Digital contraband and illegal conduct.

Contraband	Illegal Conduct
Child pornography	Possession; Receipt (18 U.S.C. §2251)
Obscene materials	Possession; Distribution (18 U.S.C. §1460)
Creative content distributed in violation of copyright laws	Copying; Distribution (18 U.S.C. §2319)
Trade secret information	Distribution (18 U.S.C. §1831)
Technology for circumventing copyright protection	Distribution (Digital Millennium Copyright Act)
Access devices (including passwords)	Possession; Distribution (18 U.S.C. §1029)

der the Computer Fraud and Abuse Act. In the Morris case, the court found that malicious intent to do damage was not an element of the crime, only the intent to access without authorization [22].

The core issue is one of "authorized access," which addresses the concerns found with the application of the concepts of trespass and invasion of privacy to computers. Criminal and civil prohibitions on trespass protect against physical intrusion or interference with property. However, prosecutions for trespass via electronic interactions with a computer lack an actual physical invasion of property. The idea of access as an element was developed for computers, and authorized access delineates permitted and non-permitted access to data and system resources.

Digital forensic analysis of the distribution of contraband material occurs in a variety of environments. These range from the examination of a local machine on a network to the active harvesting of data by a system that crawls a network, which includes node-based probing and traffic monitoring [9, 13, 16]. This paper examines the legal implications of active harvesting in which a local peer machine probes or monitors a P2P network.

2.   Basic Legal Framework

Contraband refers to artifacts or objects associated with illegal activity, which includes their possession, distribution or use as prohibited by law. State and private investigators and researchers are not immune from criminal or civil liability for examining contraband items, regardless of motive. The conduct that is illegal varies with the contraband items as described in Table 1.

Each offense may require a different level of criminal intent for guilt. However, the conduct described may be sufficient for arrest and issuance of a search warrant to search/seize any machine that may have evidence of such conduct.

Because of the range of offenses, five basic types of conduct form a framework for legal analyses relating to computer misuse. The European Convention on Cybercrime describes them as offenses against the confidentiality, integrity and availability of computer data and systems [5, 15]. The five basic types of conduct are:

- Unauthorized access to a computer, including exceeding authorized access

- Unauthorized interception of data

- Unauthorized interference with data

- Unauthorized interference with a system

- Misuse of devices

Before using a network forensic tool on a network, the impact of the tool should be evaluated in terms of these five types of conduct. A checklist consisting of the conduct and the relevant offense may be made in which the offense is identified as present, not present or unknown. For each conduct and offense, the authority or claim of right must then be identified. We demonstrate the application of this checklist on P2P research in the context of United States law, specifically 18 U.S.C. §1030 and the Computer Fraud and Abuse Act (CFAA) [11, 25].

3. Authorized Access

One may not legally access a computer or its resources without permission. This statement belies the complexity regarding what constitutes access and what constitutes authorized access. For example, an individual may have permission to access some computer resources but may not have permission to access all the resources.

It is useful to employ an analogy to physical trespass when analyzing access issues, especially when considering the need to possess certain items residing in some physical location. In an electronic environment, the possession of digital objects can take place without any physical invasion, but simply by transmitting certain commands across the network [11].

The CFAA is the primary criminal statute that addresses the unauthorized access to computers that fall under federal jurisdiction. It prohibits conduct ranging from simple (unauthorized) access to the impairment of data integrity via the transmission of hostile code.

The key elements of access and authorized access are not defined by
the CFAA. Instead, they are left to jurisprudential interpretation. A
U.S. federal court defined access as "to gain access to [or] to exercise
the freedom or ability to make use of something" in the America Online
(AOL) v. National Health Care Discount case [24]. In this case, the
defendant's emailers harvested the addresses of AOL members and sent
unsolicited bulk email to the members [24]. AOL's terms of service prohibited this particular conduct and, as such, it was deemed unauthorized
access for which AOL was entitled to civil relief.

Another court dismissed an unauthorized access charge when the evidence only showed that the defendant dialed up a computer using a
modem and viewed the login screen, but did not otherwise modify, copy
or possess anything from the computer. The appellate court noted that
until the defendant went beyond the initial banner and entered the appropriate password, there was no ability to use and, thus, no access as
commonly understood [20].

Related to the access issue is the possession of digital objects and the
use of system resources. In the case of United States v. Simpson [23],
the federal court defined possession as "the holding or having something
(material or immaterial) as one's own, or in one's control." In this
particular case, gaining root access to a remote machine gave dominion
and control and, thus, possession to all the files and resources on the
machine.

Kerr [8] notes that the technical/physical perspective is present in
other cases, as where evidence of repeated dialing activity coupled with
an admission that the conduct sought to find long-distance access codes
were found to be sufficient for access/computer trespass. Kerr also suggests that access should be defined by an analogy to physical trespass as
the making of a "virtual entrance" into a computer or from the techni-
cal/physical operation of a computing machine over a network. Under
this definition, a failed attempt to log into a machine would not be an
access, but inputting and sending data to it would constitute access.

Similarly, Madison [12] suggests that the analysis varies between "Internet as a place" (i.e., a trespass model) and "information as a thing"
(i.e., a theft model). This leaves little guidance as to what conduct constitutes access. Instead, the definition becomes subject to a requirement
of authorization or right.

"Authorized (with right)" is the express granting of permission or
right authorizing access. User agreements for online services may expressly grant access, although special terms of use may apply. States,
by statute, may authorize certain types of access to certain groups of
individuals. A court order issuing a search warrant gives the serving

officer permission to search and access a computer or system regardless of the owner's wishes. Similarly, a consent to access authorizes access just as any consent to search a physical premises obviates the need for a search warrant. The many exceptions to the requirement of a search warrant in the United States may offer other examples of *de jure* authorization (e.g., search incident to arrest), but these exceptions apply only to state law enforcement officers acting within their police powers.

The difficult issue is online activity where there is no express permission or right. There may or may not be an implicit authorization to access online systems configured for open access, which accounts for most of the content on the World Wide Web. The definition of what constitutes "implicit authorization" is a distinct issue that must be separated from conduct.

Implicit authorization is a complex issue. Hale [7] posits that the use of an open wireless local area network without express authorization is a violation of the CFAA – there is no implicit authority to use it simply due to its lack of access controls. Bierlein [2] notes that accessing an open wireless local area network is potentially a misdemeanor under the CFAA, but opines that the criminalization of such an act is unlikely. Nevertheless, such cases have been prosecuted in several jurisdictions, including the United States, Canada and Singapore [1, 10, 17].

The U.S. Court of Appeals for the First Circuit noted there could be an implicit limit on authorized access and expressly declined to adopt the view that there is a presumption of open access to Internet information [21]. The court noted that a "public website provider can easily spell out explicitly what is forbidden and, consonantly, that nothing justifies putting users at the mercy of a highly imprecise, litigation-spawning standard like reasonable expectations." Express "terms of service" have been used to delineate authorized access such that any violation of the terms becomes unauthorized access to the system [24]. However, Stanley [19] argues that access to a public web page without any violation of the terms of service should not be a CFAA violation in his analysis of the SCO Group's accusations against IBM employees who visited its website.

These examples do not clarify the legality surrounding the authority to access a computer or network where there is not an explicit permission or right. It leaves open a risk of a charge for lack of authority to which an implicit authority claim must be made as a defense. One perspective is that the Internet is based on the open exchange of information and technologies, and this inherently provides implicit authority. Alternatively, access to open systems such as wireless networks without authority has led to criminal prosecution. Perhaps, the extent of access may work along a sliding scale of access rights judged against the interference or loss to the accessed machine, data or subject of the data.

4.   Gnutella Case Study

Nasraoui, *et al.* [13] have presented a technical approach for monitoring contraband exchanges on Gnutella P2P networks. The approach [4] involves crawling through a P2P network to collect network topology data, P2P connections between nodes that identify their accessible neighbors and actual network identifiers (e.g., IP addresses of nodes). The Gnutella protocol has five messaging descriptors:

- **Ping:** A Ping is used by a machine to find other host machines that are active in a P2P network.

- **Pong:** A Pong is a response to a Ping that notes that the node is active and returns its IP address and what data it is sharing.

- **Query:** A Query is issued by a node to find out if a particular data resource is available for sharing by an active P2P node.

- **QueryHit:** A QueryHit is the positive response by a node to a Query noting that it has the data resource available for sharing.

- **Push:** A Push permits a responding machine to share its data through a firewall.

Common usage of the Gnutella protocol is to invoke an application that pings other nodes, queries for available files, and then downloads the files using the HTTP Get command.

5.   Analysis of Authorized Access

Authorized access or access with right are assumed once a machine is placed in a network and a P2P application is started without blocking the service in some way. Each Gnutella messaging descriptor makes demands on a target machine for services, responses and data. Such demands may be acceptable under an implicit authority theory for the service. Gnutella has been described as "an open, decentralized group membership and search protocol" [14], which implies permissions to participate as part of the group through open access. The Gnutella specification offers support for this feature because it describes an open exchange system that may implicitly authorize access by the very use of the protocol [4].

Table 2.   Checklist for basic legal issues in network access.

Service	Category	Machine Access	Authority or Claim of Right
Ping	Unauthorized access to a computer	No	
Query	Unauthorized access to a computer	No (Probably)	
Get	Unauthorized access to a computer	Yes	Not clear

But this assumption may be unwarranted, as indicated in the Zefer Corp. case [21], just as leaving a door unlocked does not authorize trespass into a home. Also, in the Neidorf case [6], the defendant plead guilty to unauthorized access although the access was via an open telephone dial-up modem.

Assuming a sliding scale approach is tied to system demands, the authority/right issue may be avoided for minimal message descriptors like Ping. At the other end of the spectrum, implicit authority may be required when services and data are made available (as with Get). However, this may depend on some evidence of a knowing act by the possessor of the target machine to open its services.

Ping, Query and Get constitute the bulk of the request and response services used by Gnutella P2P applications. Table 2 presents an analysis of these message descriptors in terms of authorized access.

The Ping messaging descriptor evokes a Pong response from a participating node. A Ping does not access the pinged computer as defined by statute. Rather, it makes use of a remote machine in that it provokes a response that consumes system resources (no matter how small). Note that it does not give a machine or user control over the remote machine nor does it make any machine services available for use, which is one definition of access.

When compared with the traditional notion of trespass, Ping is equivalent to knocking on a door rather than actually entering the premises, which is not considered trespass. While such analogies in computer law and technology have occasionally led litigation astray (as in the Neidorf case [6]), they can help analyze how the issue might be resolved by the courts.

The Query messaging descriptor requests information on the resources available from a particular node. A Query requires the remote machine to examine certain resources, format a response and return the requested information. A Query uses more node resources than a Ping. But it may not represent control sufficient to use the broader services of the remote machine. Using the trespass analogy, a Query is analogous to knocking on a door and asking who is behind the door without any physical entry. When viewed in terms of dominion and control of a remote system, a Query exercises minimal demands or control of system resources; it merely acquires data from the system.

The Get messaging descriptor requests a particular resource from a node and initiates an action by the remote system to return the requested resource to the originating system. Get gives the requesting machine greater control over the remote machine, permitting it to have dominion and possession of digital objects on the remote machine and (based on the configuration) to direct their copying and distribution back to the requesting machine. Thus, is the use of Get an authorized access? Probably, but a clear answer may not be possible under current jurisprudence. One possible defense is mistake of fact (e.g., I didn't know), but it would depend on the wording of the particular criminal statute (e.g., as in the Morris Internet worm prosecution [22]). In effect, this leaves the authorization for P2P crawling – and a wide array of online activities – open to interpretation.

6.   Authorized Access for P2P Research Tools

A P2P research tool can be used to harvest data about query traffic on a network [16]. The data is harvested by placing a machine/node on the network that sends a file (bit vector) to an ultrapeer. This file serves as a routing table with data asserting that the node can respond to all queries sent to the ultrapeer. The bits in the routing table are set to claim that all query keywords match files available at the node. When the ultrapeer passes queries to the node, the node stores the query data but does not respond to them.

The use of such a research tool may fall outside the expected use of and interaction with a node on a P2P network. Further, the routing table file is deceptive because it cannot respond to all queries as asserted. The tool makes a representation for the purposes of data collection, but does no file sharing.

With regard to the authorization for the tool, the context of implicit authorization would involve transactions across the network designed to facilitate its use for file and resource transfer. Because the tool acts only to consume alternative resources for purposes unrelated to the actual use of the network, it raises a question as to whether there is implicit authorization for its operation.

Note that federal statutes limit jurisdiction to situations as set out in 18 U.S.C. §1030(a)(5)(A)(B)(i) [25] causing "loss to one or more persons during any one year period ... aggregating at least $5,000 in value." Because it may be difficult or impossible to establish that this conduct constitutes a loss in the jurisdictional amount, there would seem to be no federal liability for this kind of research effort. However, for jurisdictions that do not have a financial limitation amount, there may be exposure to liability for the application of similar statutes.

7. Future Trends

The evolution of jurisprudence in this area is of importance to researchers and investigators, who should continuously monitor new statues and case decisions. Kerr [8] suggests developing new statutes related to authorized access that specifically address each type of computer misuse. A clear definition of access is required when a user sends a command to a machine that, in turn, executes the command. This rejects the virtual space/trespass analogy of a virtual entry into a machine. However, Kerr proposes that this broad meaning of access would change the definition of access without authorization to access that circumvents restrictions by code. This negates contract law, keeps out analogs and, as with the Digital Millennium Copyright Act, offers regulation to control access and ensure security.

The evolution of jurisprudence will also be affected as more case law develops regarding other types of contraband (i.e., not restricted to child pornography). The jurisprudence related to copyright prosecutions of operators of P2P nodes may come to treat such activity differently. Similarly, laws governing privacy rights may impact this analysis by focusing further on an invasion of rights rather than a physical machine. Simon [18], who describes a possible continuum of online privacy expectations, posits that gaining evidence from a public chat room would not violate the Fourth Amendment, but the use of services that increasingly protect privacy and control may render an online search in violation of the Fourth Amendment or the Electronic Communications Privacy Act.

It may also be necessary to consider whether legislative changes are needed to provide opportunities for open research. There is a legislative "safe harbor" for lawfully-authorized investigative activity by a law enforcement agency under the U.S. computer access statute, but this does not apply to private investigators and researchers. Indeed, closing off open research and limiting research to law enforcement entities may be handing an advantage to those using P2P networks for illegal purposes, just as earlier federal limitations on encryption research damaged such efforts within the United States.

8.   Conclusions

Several pitfalls exist concerning the analysis of system use in networks, both for investigators and researchers. Continued analysis of the legal and ethical implications of the techniques is important to performing investigations as well as developing and testing forensic tools.

Caloyannides [3] observes that, where digital evidence is concerned, "the potential for a miscarriage of justice is vast." It is essential that any network research that seeks to address misconduct in the use of networks – whether of contraband transactions or other illegal activity – take into account possible legal restrictions. Good intentions are not enough. Failure to satisfy the legal constraints can compromise the evidentiary value of investigations as well as expose investigators and researchers to legal liability and damage to reputation.

References

[1] Associated Press, Singapore teen faces 3 years' jail for tapping into another's wireless Internet, *International Herald Tribune*, November 10, 2006.

[2] M. Bierlein, Policing the wireless world: Access liability in the open Wi-Fi era, *Ohio State Law Journal*, vol. 67(5), pp. 1123–1186, 2006.

[3] M. Caloyannides, *Privacy Protection and Computer Forensics*, Artech House, Norwood, Massachusetts, 2004.

[4] Clip2 Distributed Search Services, The Gnutella Protocol Specification v0.4 (Document Revision 1.2) (www9.limewire.com/developer /gnutella_protocol_0.4.pdf), 2001.

[5] Council of Europe, Convention on Cybercrime, Strasbourg, France (conventions.coe.int/Treaty/en/Treaties/Html/185.htm), 2001.

[6] M. Godwin, Some "property" problems in a computer crime prosecution, Electronic Frontier Foundation, San Francisco, California (www.textfiles.com/law/cardozo.txt), 1992.

[7] R. Hale, Wi-Fi liability: Potential legal risks in accessing and operating wireless Internet, *Santa Clara Computer and High Technology Law Journal*, vol. 21, pp. 543–559, 2005.

[8] O. Kerr, Cybercrime's scope: Interpreting "access" and "authorization" in computer misuse statutes, *NYU Law Review*, vol. 78(5), pp. 1596–1668, 2003.

[9] S. Kwok, P2P searching trends: 2002-2004, *Information Processing and Management*, vol. 42(1), pp. 237–247, 2006.

[10] A. Leary, Wi-Fi cloaks a new breed of intruder, *St. Petersburg Times*, July 4, 2005.

[11] M. Losavio, The law of possession of digital objects: Dominion and control issues for digital forensic investigations and prosecutions, *Proceedings of the First International Workshop on Systematic Approaches to Digital Forensic Engineering*, pp. 177–183, 2005.

[12] M. Madison, Rights of access and the shape of the Internet, *Boston College Law Review*, pp. 433–507, 2003.

[13] O. Nasraoui, D. Keeling, A. Elmaghraby, G. Higgins and M. Losavio, Node-based probing and monitoring to investigate the use of peer-to-peer technologies for the distribution of contraband material, *Proceedings of the Third International Workshop on Systematic Approaches to Digital Forensic Engineering*, pp. 135–140, 2008.

[14] M. Ripeanu, A. Iamnitchi and I. Foster, Mapping the Gnutella network, *IEEE Internet Computing*, vol. 6(1), pp. 50–57, 2002.

[15] S. Schjolberg, The legal framework – Unauthorized access to computer systems, Penal legislation in 44 countries, Moss District Court, Moss, Norway (www.mosstingrett.no/info/legal.html), 2003.

[16] S. Sharma, L. Nguyen and D. Jia, IR-Wire: A research tool for P2P information retrieval, *Proceedings of the Second Workshop on Open Source Information Retrieval*, pp. 33–38, 2006.

[17] R. Shim, Wi-Fi arrest highlights security dangers, *ZDNet News*, November 28, 2003.

[18] B. Simon, Note: The tangled web we weave – The Internet and standing under the Fourth Amendment, *Nova Law Review*, vol. 21, pp. 941–959, 1997.

[19] J. Stanley, Whose hands are "unclean?" – SCO, IBM's "agents" and the CFAA, *Groklaw* (www.groklaw.net/articlebasic.php?story =20041217091956894), December 17, 2004.

[20] Supreme Court of Kansas, State of Kansas v. Allen, *Pacific Reporter (Second Series)*, vol. 917, pp. 848–854, 1996.

[21] U.S. Court of Appeals (First Circuit), EF Cultural Travel BV v. Zefer Corp., *Federal Reporter (Third Series)*, vol. 318, pp. 58–64, 2003.

[22] U.S. Court of Appeals (Second Circuit), United States v. Morris, *Federal Reporter (Second Series)*, vol. 928, pp. 504–512, 1991.

[23] U.S. Court of Appeals (Tenth Circuit), United States v. Simpson, *Federal Reporter (Third Series)*, vol. 94, pp. 1373–1382, 1996.

[24] U.S. District Court (Northern District of Iowa, Western Division), America Online, Inc. v. National Health Care Discount, Inc., *Federal Supplementer (Second Series)*, vol. 121, pp. 1255–1280, 2001.

[25] U.S. Government, Title 18, Crimes and Criminal Procedure, *United States Code Annotated*, Washington, DC, pp. 445–453, 2000.

Chapter 20

A SURVEY OF THE LEGAL ISSUES FACING DIGITAL FORENSIC EXPERTS

Sydney Liles, Marcus Rogers and Marianne Hoebich

Abstract   This paper discusses the results of a survey focusing on the legal issues facing digital forensic experts in the United States. The survey attracted 71 respondents from law enforcement, academia, government, industry and the legal community. It extends the well-known Brungs-Jamieson research on attitudes and priorities of the Australian digital forensic community. The results are compared with those from the Brungs-Jamieson study to determine if digital forensic experts from different countries share priorities and concerns. Several differences are observed between stakeholder groups regarding the importance of specific legal issues. Nevertheless, the results indicate that, despite differing opinions, it is possible to find a common ground that can help craft public policy and set funding priorities.

Keywords: Legal issues, digital forensic experts, survey

1.   Introduction

The primary purpose of digital forensics is to present digital evidence in legal proceedings. Therefore, the techniques employed to extract digital evidence from devices must comply with legal standards. However, due to the nature of the Internet, digital forensic investigations are not constrained by geographical boundaries and legal issues are complicated by the presence of multiple jurisdictions.

An electronic crime initiated in Australia can bring down a computer system in the United States (or vice versa). Consequently, it is important that there is a cohesive movement towards the acceptance of legal standards for digital evidence in international courts of law.

Jurisdictional issues are among the most common problems reported in the literature [2, 6, 10, 11]. Because cyber crime is not constrained

G. Peterson and S. Shenoi (Eds.): Advances in Digital Forensics V, IFIP AICT 306, pp. 267–276, 2009.
© IFIP International Federation for Information Processing 2009

by territorial, state or national boundaries, there are often questions about the jurisdiction where the crime occurred and the agency with the authority to investigate and prosecute. International cooperation is a related issue – a cyber crime can occur anywhere in the world, have victims in different locations and leave trails of evidence that cross multiple national boundaries. The need to enact cyber crime laws on an international scale is an ongoing effort as is the need to improve cooperation among countries [10, 11]. Most researchers agree that new laws are probably not required as most nations and states have cyber crime laws. However, existing laws need improved definitions and clarification on several important points [2, 6, 10, 11].

Brungs and Jamieson [4] conducted research on the attitudes and priorities of the Australian digital forensic community. Their study, which identified seventeen legal issues in three categories (judicial, privacy and multi-jurisdictional), laid the groundwork for classifying legal issues related to digital forensics.

The Brungs-Jamieson study covered Australian telecommunications legislation, namely the Telecommunications Act of 1979 and its interpretation. The study identified the need to protect the privacy of individuals and businesses during investigations as a major challenge. Other researchers [7, 8, 11, 14] have also noted that this is a major issue in digital forensics.

The presentation of digital evidence in legal proceedings is another important issue. Because lawyers, judges and juries may have limited technical knowledge, the presentation of digital evidence must be done in a clear, easily understandable manner [3, 5, 8, 14]. Broucek and Turner [3] note that most legal professionals have a limited understanding of technology and tend to lack confidence in the ability of technical specialists to produce evidence that is admissible in a court of law.

Related work confirms the issues raised by Brungs and Jamieson concerning best practices, testing of digital forensic tools and expert witnesses. Numerous digital forensic techniques are used by investigators and examiners; however, no best practice guides are currently available. Also, there currently are no published error rates or testing results for digital forensic tools [5, 9, 12, 13]. The qualifications and skills of expert witnesses is also a serious issue. Meyers and Rogers [9] question whether one can be considered an expert based on the ability to use a tool or software package, but without the ability to clearly define how the tool works or without reviewing the source code. Attempts are underway to develop standards for expert qualifications [1, 12], but none exist at present.

Brungs and Jamieson identified many significant legal issues facing the discipline of digital forensics. However, while much has been written about the individual issues, little has been done to clarify the issues or to determine where the digital forensic community should focus its efforts. This study explores the same legal issues as Brungs and Jamieson, but in the context of the U.S. digital forensic community.

2.   Brungs-Jamieson Survey

The Brungs-Jamieson study surveyed the attitudes and priorities of digital forensic experts in Australia. It identified seventeen key legal issues, which were divided into three categories: judicial, privacy and multi-jurisdictional. The study laid the groundwork for the classification of legal issues and the creation of a taxonomy. However, it appears that no follow-up research has been conducted related to the Brungs-Jamieson survey.

Brungs and Jamieson set out to accomplish two goals: (i) identify a set of legal issues facing digital forensics, and (ii) determine the importance of the identified issues to three stakeholder groups: police, regulators and consultants. A Delphi methodology was used to survey a panel of eleven Australian experts in order to identify the principal legal issues. After identifying seventeen issues, the experts were asked to rank them from 1 (highest priority) to 17 (lowest priority), and to rate each issue on a seven-point Likert scale from 1 (unimportant) to 7 (very important). All the issues were rated 3 or lower on an inverted scale from 1 (very important) to 7 (unimportant). The top five issues were "Jurisdictional," "Telecommunications Act covering data, "Interpretation of Telecommunications Act," "International cooperation in practice," and "Revision of mutual assistance." High concordance was observed between the importance ratings and average rankings, which was confirmed using a Kendall's W statistical test ($W = 0.974$, $p = 0.013$) [4].

The Brungs-Jamieson study also reported the average rankings of each issue by group. However, the method for determining this ranking was not reported. The average rankings were converted to ranks from 1 to 17 for comparison across groups.

3.   Survey Methodology and Results

This study builds on the Brungs-Jamieson research by conducting a similar survey of digital forensic experts in the United States. The respondents included law enforcement, academics, government, industry and legal experts. The seventeen issues identified by Brungs and Jamieson were used to confirm and refine an initial taxonomy.

Our study involved a voluntary, anonymous, self-selecting web-based survey of digital forensic experts. The following issues were presented to the survey participants:

- **Issue 1:** Jurisdictional (state to state and federal to state)

- **Issue 2:** Computer evidence presentation difficulties

- **Issue 3:** Criminal prosecution vs. civil litigation

- **Issue 4:** International cooperation in legal practice

- **Issue 5:** Access and exchange of information

- **Issue 6:** Confidential records and business systems privacy

- **Issue 7:** Privacy protection for data transmission laws

- **Issue 8:** Privacy issues and workplace surveillance

- **Issue 9:** Interpretation of laws affecting digital evidence

- **Issue 10:** Preservation of privacy of clients during digital investigations

- **Issue 11:** Launching actions against persons unknown in civil litigation

- **Issue 12:** Requirement for best practices guides and standards

- **Issue 13:** Computer literacy in the legal sector

- **Issue 14:** Contrast of broadcast vs. communications

- **Issue 15:** Need to specify new offenses

- **Issue 16:** Testing of new tools and techniques

- **Issue 17:** Expert witness skills and qualifications

The respondents were asked to rank each of the seventeen issues using a five-point Likert Scale ranging from 1 (not important) to 5 (most important). The survey was accessed from a web page hosted by the Center for Education and Research in Information Assurance and Security (CERIAS) at Purdue University from October 26, 2007 to November 20, 2007.

The survey was promoted by sending invitations to digital forensic professionals from around the United States. Emails were sent to authors of published research papers related to digital forensics and the law. A link to the survey was also posted on technical forums on the Internet. Additionally, calls for participation were sent to companies, government agencies and universities with strong interests in information assurance and digital forensics.

A total of 71 respondents completed the online survey. The respondents were from law enforcement ($n = 13$), academia ($n = 26$), government ($n = 9$), legal/courts ($n = 3$) and commercial ($n = 20$).

Prior to analysis, the data was examined for accuracy, missing entries and the satisfaction of the assumptions for performing multivariate analysis. The data had no missing values. However, the answers provided by two respondents were found to be univariate outliers for six of the seventeen issues. Further examination revealed that all the answers provided by these two respondents had extreme values. The data provided by these two respondents was eliminated, leaving 69 responses for the final analysis.

In addition, Issue 11 (Launching actions against persons unknown in civil litigation) showed numerous outliers, which indicated considerable confusion among respondents about this issue. Issue 11 was therefore eliminated from further analysis.

Each of the remaining sixteen issues was treated as a variable of interest in the data analysis. Examination of skewness and kurtosis, the application of the Kolmogorov-Smirnov and Shapiro-Wilk tests, and visual inspection of histograms, box plots and Q-Q plots verified that the data was not normally distributed for any of the sixteen issues. Therefore, the data was analyzed using non-parametric statistical tests.

3.1 Survey Results

The Pearson and the Spearman correlation tests showed a significant correlation between Issue 4 (International cooperation in legal practice) and Issue 5 (Access and exchange of information) by group. In particular, Issue 4 had $r(67) = -0.320$, $p < 0.01$ (two-tailed) and $rs(67) = -0.334$, $p < 0.01$ (two-tailed). Issue 5 had $r(67) = 0.320$, $p < 0.01$ (two-tailed) and $rs(67) = 0.299$, $p < 0.05$ (two-tailed) by group.

A Kruskal-Wallis test was performed to determine the mean ranking of each issue by group. The results of the test (mean rankings) are shown in Table 1. A higher number indicates a higher ranking or greater importance as identified by the group. Note that Group 2 denotes law enforcement ($n = 13$), Group 3 denotes academia ($n = 25$), Group 4 denotes government ($n = 9$), Group 5 denotes legal/courts ($n = 3$) and Group 6 denotes commercial entities ($n = 19$).

Table 1.   Kruskal-Wallis test results.

Issue		Group				
		2	3	4	5	6
1	Jurisdictional	41.15	31.64	43.83	21.83	33.11
2	Presentation Difficulties	32.88	40.30	41.39	12.00	20.08
3	Criminal vs. Civil	31.15	38.00	32.28	28.67	35.97
4	International Cooperation	42.04	39.24	33.11	39.33	24.82
5	Access and Exchange Information	29.46	32.66	24.61	34.00	46.95
6	Confidential Records	31.31	32.88	33.67	48.33	38.84
7	Data Transmission Privacy	41.46	33.50	32.72	41.00	32.68
8	Work Surveillance	32.92	38.80	40.67	40.67	27.84
9	Interpretation of Laws	33.54	33.12	40.67	60.00	31.84
10	Client Privacy	27.31	40.60	26.83	37.50	36.37
12	Best Practices	43.08	34.94	25.11	41.67	33.18
13	Literacy in Legal Sector	32.31	37.40	27.28	57.50	33.79
14	Broadcast vs. Communications	33.85	41.24	34.67	7.00	32.16
15	New Offenses	38.15	33.72	34.44	19.33	27.36
16	Testing of New Tools	31.19	35.86	37.94	22.33	37.08
17	Expert Witness	27.04	38.66	34.33	38.17	35.45

3.2   Analysis of Results

In order to permit a comparison with the Brungs-Jamieson results, the data was converted into a separated data set with scores ranging from 1 (very important) to 5 (unimportant). Kendall's W test was performed for the three groups (law enforcement, government and commercial) that were comparable to the Brungs-Jamieson groups. A one-to-one comparison of results was not possible because our study included two additional groups (academia and legal/courts), which are also legitimate stakeholders. Results corresponding to these additional groups will be included in future reports.

Table 2 compares the results of Kendall's W tests for our survey and the Brungs-Jamieson survey for the three common groups (law enforcement, government and commercial). Note that the non-parenthesized values in the table represent mean rankings while the values in parentheses correspond to issue rankings.

The results indicate that differences exist in the Kendall's W rankings for the two surveys. Both the actual values and the rankings show differences between groups. However, it is interesting to note that in both studies the law enforcement group ranked the need to specify new offenses fairly low (Rank 14 in our study and Rank 12 in the Brungs-Jamieson study). Also, the need for international cooperation

Table 2. Comparison of Kendall's W test results.

Issue	Law Enforcement		Government		Commercial	
	Liles	B-J	Liles	B-J	Liles	B-J
Jurisdictional	7.88(8)	7.00(5)	7.00(5)	4.75(3)	9.53(11)	8.67(7)
Presentation Difficulties	8.50(9)	5.67(3)	6.33(3)	5.50(4)	9.03(10)	9.00(9)
Criminal vs. Civil	11.50(15)	10.67(14)	11.78(15)	13.00(16)	10.34(12)	7.00(5)
International Cooperation	6.85(4)	5.33(2)	7.72(9)	3.25(1)	10.55(14)	9.67(10)
Access and Exchange Information	9.27(11)	6.00(4)	10.94(13)	6.75(7)	5.26(1)	12.33(14)
Confidential Records	7.08(5)	10.33(12)	6.67(4)	8.00(8)	5.68(2)	13.00(17)
Data Transmission Privacy	5.23(2)	10.00(10)	7.56(7)	6.50(6)	7.08(5)	6.00(3)
Work Surveillance	9.23(10)	7.33(7)	7.50(6)	10.25(11)	10.53(13)	4.00(1)
Interpretation of Laws	6.73(3)	4.33(1)	4.89(1)	4.25(2)	7.34(6)	5.00(2)
Client Privacy	9.96(12)	10.00(10)	10.06(12)	6.00(5)	7.92(8)	8.67(7)
Best Practices	5.08(1)	11.00(15)	9.33(11)	11.50(13)	7.42(7)	12.67(15)
Literacy in Legal Sector	7.42(7)	9.33(9)	8.94(10)	12.25(14)	7.03(4)	10.00(11)
Broadcast vs. Communications	12.50(16)	7.00(5)	12.33(16)	10.50(12)	12.92(16)	10.67(13)
New Offenses	11.46(14)	10.33(12)	11.33(14)	8.50(9)	11.11(15)	12.67(15)
Testing of New Tools	7.35(6)	8.67(8)	6.00(2)	10.00(10)	6.26(3)	6.67(4)
Expert Witness	9.96(13)	3.33(16)	7.61(8)	12.75(15)	8.00(9)	7.00(5)

was ranked fairly high (Rank 4 in our study and Rank 2 in the Brungs-Jamieson study).

The government groups in both studies gave high rankings to the interpretation of laws, presentation difficulties and jurisdictional issues. However, there was no agreement between the government groups regarding the issues that received low rankings.

On the other hand, the commercial groups in the two studies found some common agreement on the need to test new tools and to protect client privacy; they also agreed on low rankings for new offenses. Nevertheless, it is interesting to note that there is little, if any, agreement across groups regarding the importance of the sixteen issues.

4.   Discussion

The results of the current study do indeed differ from those of the Brungs-Jamieson study. Unfortunately, the Brungs-Jamieson data set is not available (and it may not be detailed enough), so it is not possible to determine the factors responsible for the differences. Two possible reasons are the differing sizes of the data sets ($N = 69$ for the current data set while $N = 11$ for the Brungs-Jamieson data set) and the fact that the surveys were conducted in different countries. But these are mere speculation and additional research is required to fully explore this question.

The current study indicates marked differences between stakeholder groups regarding the rankings and, therefore, the importance of the sixteen legal issues. Based on the Kruskal-Wallis test results, the law enforcement group ranked best practices as the most important issue while the government group rated jurisdictional issues and the commercial group ranked access and exchange of information as the most important. This trend holds for the second and third ranked issues for each group. Law enforcement ranked international cooperation as the second most important issue while the government group ranked presentation difficulties and the commercial group ranked confidential records and business systems privacy as the second most important issue. The third ranked issues are privacy protection for data transmission laws in the case of the law enforcement group, privacy issues and workplace surveillance for the government group and the need to specify new offenses for the commercial group.

The rankings of two issues showed agreement across groups. The law enforcement and government groups ranked Issue 14 (Contrast of broadcast vs. communications) as the sixth most important issue; the commercial group ranked this issue twelfth. Issue 7 (Privacy protection for data transmission laws) was ranked eleventh by the government and commercial groups, and third by the law enforcement group.

While some of the results differ from those of the Brungs-Jamieson study, the two studies share a common finding – stakeholder groups disagree on the importance of specific legal issues. This is expected because digital forensics is an interdisciplinary field with multiple stakeholder groups, each with different priorities regarding the legal issues.

5.   Conclusions

Despite the exploratory nature of the survey and limitations in research design, the finding that law enforcement, government and commercial experts disagree on the importance of specific legal issues that face digital forensics is significant. In order to have effective governance and allocate limited resources, it is important to understand the priorities of all the principal stakeholders in the discipline of digital forensics. The study also suggests that, while the stakeholders disagree about the individual issues, it may be possible to find common ground if the issues are examined more broadly. For example, the top issues in this study (international cooperation, jurisdiction and access and exchange of information) should be examined for areas of overlap rather than just the differences. Identifying the common areas can assist in crafting public policy and in setting funding priorities.

References

[1] V. Baryamureeba and F. Tushabe, The enhanced digital investigation process model, *Proceedings of the Fourth Digital Forensic Research Workshop*, 2004.

[2] R. Broadhurst, Developments in the global law enforcement of cyber-crime, *Policing: International Journal of Police Strategies and Management*, vol. 29(3), pp. 408–433, 2006.

[3] V. Broucek and P. Turner, Bridging the divide: Rising awareness of forensic issues amongst systems administrators, presented at the *Third International System Administration and Networking Conference*, 2002.

[4] A. Brungs and R. Jamieson, Identification of legal issues for computer forensics, *Information Systems Management*, vol. 22(2), pp. 57–66, 2005.

[5] M. Carney and M. Rogers, The Trojan made me do it: A first step in statistical based computer forensics event reconstruction, *International Journal of Digital Evidence*, vol. 2(4), 2004.

[6] J. Conley and R. Bryan, A survey of computer crime legislation in the United States, *Information and Communications Technology Law*, vol. 8(1), pp. 35–58, 1999.

[7] N. King, Electronic monitoring to promote national security impacts workplace privacy, *Employee Responsibilities and Rights Journal*, vol. 15(3), pp. 127–147, 2003.

[8] R. Laubscher, D. Rabe, M. Olivier, J. Eloff and H. Venter, Computer forensics for a computer-based assessment: The preparation phase, *Proceedings of the Fifth Annual Information Security South Africa Conference*, 2005.

[9] M. Meyers and M. Rogers, Computer forensics: The need for standardization and certification, *International Journal of Digital Evidence*, vol. 3(2), 2004.

[10] F. Pocar, New challenges for international rules against cyber-crime, *European Journal on Criminal Policy and Research*, vol. 10(1), pp. 27–37, 2004.

[11] L. Reid, Expert opinion: Interview with Amanda M. Hubbard, J.D., Fulbright Scholar, former trial attorney, Computer Crime and Intellectual Property Section, U.S. Department of Justice, *Journal of Information Privacy and Security*, vol. 2(1), pp. 47–56, 2006.

[12] M. Reith, C. Carr and G. Gunsch, An examination of digital forensic models, *International Journal of Digital Evidence*, vol. 1(3), 2002.

[13] M. Saudi, An Overview of Disk Imaging Tools in Computer Forensics, InfoSec Reading Room, SANS Institute, Bethesda, Maryland, 2001.

[14] F. Witter, Legal Aspects of Collecting and Preserving Computer Forensic Evidence, InfoSec Reading Room, SANS Institute, Bethesda, Maryland, 2001.

Chapter 21

AN EXTENDED MODEL FOR E-DISCOVERY OPERATIONS

David Billard

Abstract   Most models created for electronic discovery (e-discovery) in legal proceedings tend to ignore the technical aspects mainly because they assume that only traditional digital forensic tasks are involved. However,
this assumption is incorrect. The time frames for conducting e-discovery
procedures are very restricted, and investigations are carried out in real
time with strict non-disclosure dispositions and changing demands as
the cases unfold. This paper presents an augmented model and architecture for e-discovery designed to cope with the technological complexities in real-world scenarios. It also discusses how e-discovery operations
should be handled to ensure cooperation between digital forensic professionals and legal teams while guaranteeing that non-disclosure agreements and information confidentiality are preserved.

Keywords: Electronic discovery, technical aspects, non-disclosure

1.   Introduction

Electronic discovery (e-discovery) refers to any process in which electronic data is sought, located, secured and searched with the intent of
using it as evidence in civil or criminal legal proceedings [3]. The most
popular e-discovery model is the Electronic Discovery Reference Model
(EDRM) [2], which is presented in Figure 1.

EDRM expresses the phases of e-discovery from the point of view
of an attorney. The six e-discovery phases, which are very similar to
those proposed by McKemmish [4] for digital forensic investigations, are
summarized as:

- **Information Management:** This phase is not necessarily part
  of e-discovery. Rather, it is a pre-processing step that should be
  performed by an entity in case litigation should occur.

G. Peterson and S. Shenoi (Eds.): Advances in Digital Forensics V, IFIP AICT 306, pp. 277–287, 2009.

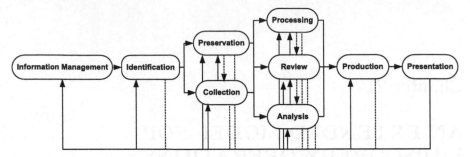

Figure 1. Electronic Discovery Reference Model.

- **Identification:** This phase involves the location of the potential evidence containers.

- **Preservation and Collection:** This phase deals with the preservation and collection of the potential evidence containers, during which exact copies of the evidence are made in a forensically-sound manner. Note that only the evidence deemed to be relevant to the case is considered.

- **Processing, Review and Analysis:** This phase involves the processing of evidence by digital forensic experts. The results of the processing are reviewed by legal teams and analyzed by the concerned parties.

- **Production:** This phase involves the production of privileged material in a human-readable format.

- **Presentation:** This phase involves the presentation of the evidence in the legal framework of the case.

Unfortunately, EDRM is too simplistic to capture the complexities of e-discovery in the real world. For example, the Processing, Review and Analysis Phase often involves several (possibly hundreds of) individuals with different qualifications and roles operating under different legal and contractual frameworks. Moreover, the information that is ultimately produced (i.e., discovered) along with its byproducts are often designated to be released to certain individuals and not others.

EDRM is also unnecessarily complicated by the presence of many return paths (arrows in Figure 1). Some of the return paths do not need to exist and may even introduce flaws in e-discovery procedures. Consequently, we propose to augment EDRM and, at the same time, eliminate certain return paths.

E-discovery processes are tightly controlled by procedures and court orders and usually operate in restricted and regulated time frames. For example, Rule 26(a) of the U.S. Federal Rules of Civil Procedure allows for an initial disclosure before an actual discovery request is made; Rule 16(b) imposes a scheduling order; and Rule 26(f) requires the parties to confer at least 21 days before a scheduling order is due. Fortunately, even if each e-discovery case is unique, it is possible to capitalize on certain invariants.

The paper presents an augmented model for e-discovery. It identifies the various actors involved in e-discovery and their roles, and proposes an augmentation to EDRM designed to cope with the technological complexities in real-world scenarios. Finally, it discusses how e-discovery processes should be handled to ensure cooperation between digital forensic professionals and legal teams.

2. E-Discovery Actors

Several individuals and teams of individuals are involved in e-discovery operations. These actors do not have the same levels of knowledge about the case and are bound by various contracts and non-disclosure agreements. We distinguish four actors that operate with respect to these non-disclosure agreements.

- **Digital Forensic Team:** This team is responsible for extracting and collecting potential evidence from all types of devices: hard drives, cell phones, backup tapes, GPS devices, etc. The potential evidence containers are carved, decrypted, de-duplicated, indexed, searched and made user-readable using advanced forensic tools and dedicated software.

- **Research Teams:** These teams usually comprise attorneys and legal assistants who have access to the potential evidence. The role of the research teams is to pre-sort information pertaining to the case as "privileged" (i.e., information pertinent to the resolution of the case), "confidential" (i.e., private information about the individuals whose devices were searched), and "irrelevant." In general, there are at least two research teams, one for each party involved in an e-discovery case. The research teams focus on the meaning of the documents and, therefore, may not incorporate technical personnel.

- **Party Counsels:** These actors are the attorneys who are in charge of procedures on behalf of their respective clients (parties). They have the ultimate say about relevant information pertaining

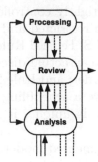

Figure 2.   Processing, Review and Analysis Phase.

to the case. They guide the research teams' activities and may ask
for additional investigations to be conducted by the digital forensic
team.

- **Chief of Forensic Operations:** This individual is in charge of
  dispatching evidence to the research teams; maintaining the technological means to ensure non-disclosure; securing privileged information along with the scientific processes that support its findings; and liaising with the digital forensic team, research teams
  and party counsels to ensure that the processes are carried out
  correctly. Needless to say, serving as a chief of forensic operations
  is a most demanding task, with intense pressure and a close relationship to the case core.

The interactions between these four actors are complex and vary con-
siderably throughout the e-discovery process. This complexity is captured using the e-discovery framework described in the next section.

3.   E-Discovery Framework

The framework described in this section is intended to fully support
e-discovery processes in the real world. The first three EDRM phases
are relatively traditional. They do not involve research teams and, from
the technical point of view, can be handled adequately by trained digital
forensic professionals using state-of-the-art tools. However, the fourth
EDRM phase, Processing, Review and Analysis, is complex and requires
special consideration.

3.1   Processing, Review and Analysis Phase

The Processing, Review and Analysis Phase is illustrated in Figure
2. This phase of e-discovery is the most complex and costly. In the

following sections, we discuss the processing, review and analysis steps in detail and associate them with the various actors involved in e-discovery operations.

Processing: The processing step involves several tasks.

- **Document Carving:** Carving is used to retrieve documents, images, audio, video and, above all, email. Several forensic tools are available to accomplish this step; they can retrieve deleted documents as well as documents embedded in emails, compressed files and archives.

- **Decryption:** Some organizations use cryptography to secure their data. This is usually a good policy and is recommended in the normal course of business. Unfortunately, it complicates the work of the digital forensic team because decryption keys and passphrases may be missing.

- **De-Duplication:** When dealing with a corporate email system or document repository, it is often the case that the same document is found multiple times. By "same document" we mean a document with the same contents and metadata. For example, a document emailed by an executive to company employees may be present in the sender's mailbox and in the mailboxes of the other recipients, and even more mailboxes if the original mail was forwarded to others. While the fact that the document was sent might be important to the case, it is unnecessary (and a waste of resources) to preserve every copy of the email during e-discovery. Consequently, the digital forensic team would use de-duplication tools to identify duplicate files and retain only one copy of the file.

- **Search Indexing:** In general, e-discovery operations rely heavily on keyword searches. Consequently, it is important to index data and to use powerful search engines.

- **Presentation:** The purpose of the processing step is to create content for the research teams. The content must be delivered in a format that enables the research teams to sort and label the documents quickly and efficiently. Documents should be pre-categorized with respect to their potential value to the case. For instance, documents written in French might be relevant and those in German less relevant, or emails with attached spreadsheets should be examined first.

- **Comparison Chart/Timeline Preparation:** File content is not the only information of interest in e-discovery operations. For example, it might be important to know if a person engaged in certain stock market transactions before or after receiving an email. In such an instance, the digital forensic team has to create charts and timelines from the available files and their metadata, and from information pertaining to the files (e.g., dates). Investigative tools are available to facilitate the production of charts and timelines.

- **Repository Creation:** Increasing amounts of information are maintained in databases as part of a software suite or in databases built specifically for organizations. It is of extreme importance to be able to connect to and access information from these databases without disrupting normal business operations. The collected information could be stored in a specially-designed repository and searched using business intelligence tools. These tools sort through the data, present information in a condensed form and offer analytic services.

The seven tasks described above are performed by the digital forensic team with input from a party counsel. The party counsel would provide the categories of documents and the keywords to be used in searches. Typically, there is close cooperation between the forensic team and the party counsel in developing the keyword list. For example, the party counsel might provide a list of nicknames or a list of phone numbers to be used in searches.

Review: The review step involves the examination of the documents produced during the processing step. The documents to be examined are given to the research teams by the party counsels (via the digital forensic team) as the case goes along. The party counsels orchestrate the information flow during the review step.

It is important to prevent information leaks during the review step. The individuals participating in the review step can be in the presence of very sensitive documents, including documents that are not pertinent to the case. Therefore, the research teams must work in strictly-controlled physical locations with no telephone service or Internet connectivity. Computers used by the research teams should have their USB, firewire, wireless and CD/DVD functionality disabled. Also, all internal network communications should be encrypted.

The research teams sort the documents based on the filters provided by the party counsels. As a result, the documents are categorized as "privileged" (relevant to the case), "confidential" (private nature), or "irrelevant" (not processed any further).

Analysis: The party counsels analyze the privileged documents with respect to the case objectives and applicable laws. The analysis could lead to additional processing as well as the inclusion of new evidence containers. For example, the analysis of the documents might shed light on the behavior of a new person in the case and his cell phone becomes a new potential evidence container. The party counsels may also request the chief of forensic operations to conduct new searches. The chief of forensic operations quantifies the duration of the searches, oversees their progress and ensures that the schedule is maintained.

3.2   Additional Steps

The additional steps include the documentation of forensic processes and the cleaning of digital media.

Forensic Process Documentation: Every forensic task performed during the Processing, Review and Analysis Phase should be documented in detail. Documentation may not be considered as a step as such, but it should be done continuously.

Digital Media Cleaning: All digital media should be cleaned at the end of the Processing, Review and Analysis Phase. This includes every computer used by the digital forensic team, research teams and chief of forensic operations that could have any data relating to the case. Two common cleaning techniques are to erase all data using a DoD-certified method or to physically destroy all the data containers. The first method is very time consuming – several hours may be required to wipe a single hard drive. The second method, which we believe is more appropriate, is to destroy the data containers using hammers, drilling tools and possibly fire and acid.

3.3   Modified E-Discovery Reference Model

We propose a modified EDRM incorporating both simplification and augmentation. The EDRM workflow is simplified by incorporating fewer feedback loops:

- The party counsels might identify a new potential evidence container. In this case, the potential evidence container must be treated as new evidence and should be preserved.

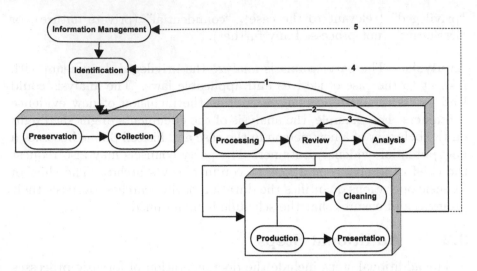

Figure 3.   Modified EDRM.

- The party counsels may alter the keyword lists or search filters. In this case, the new information is sent to the digital forensic team for processing.

- The party counsels might modify their reviewing criteria. In this case, the research teams are informed about the change.

- Additional data may be sought by the judge or by the parties after the privileged information is produced and presented. In this case, a new identification phase is initiated.

- The lessons learned during the entire process are integrated in corporate information management systems in the event of additional e-discovery demands.

We also introduce a cleaning step to the model in order to erase all the data on the devices used in e-discovery.

Figure 3 presents our modified EDRM schema. The schema is simpler, but more accurately reflects the complexity of the e-discovery process. The boxes represent atomic, closed steps corresponding to distinct units of work. (Note that "production," "cleaning" and "presentation" fall in the same unit of work.) Some additional return paths can be drawn; however, we believe they constitute exceptions and are, therefore, not included.

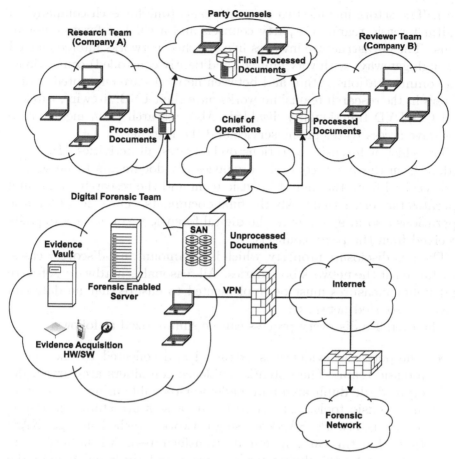

Figure 4. Technical architecture for e-discovery.

3.4   Technical Architecture

We have designed an architecture that supports our modified EDRM (Figure 4). This architecture has been used in a real case involving companies in the United States and Europe.

Note, however, that every case is unique and the specific e-discovery setup may have to be altered to match the objectives and local support. Sometimes, a setup has to be reproduced. For example, in a multinational case, collected data may not be transferred legally from one country to another because of different laws [1]. Therefore, the setup has to be reproduced in each country and the chief of forensic operations at one of the sites serves as the "chief of global operations."

Our example in Figure 4 has two opposing companies that are processing potential evidence (emails, spreadsheets, mobile phone calendars, etc.). The actors include two research teams (one for each company), a digital forensic team, two party counsels and a chief of forensic operations. The infrastructure involves five switched networks interconnected through gateways with very limited (and tightly controlled) connections. All communications within and between networks are encrypted. Computers in the research teams' networks have their USB, firewire, wireless and CD/DVD functionality disabled. Also, research team members do not have access to telephone service and the Internet.

The chief of forensic operations orchestrates all activities. He takes orders from the party counsels, organizes the documents to be sent to and received from the digital forensic team and the research teams, and operates the server that hosts the final documents. The chief of forensic operations also interacts with the digital forensic team on new requests received from the party counsels.

The non-disclosure property, which is paramount in e-discovery cases, is achieved at the network boundaries. All reasonable hardware, software and policy measures must be implemented to ensure that no data can leave the secured networks.

The overall e-discovery process can be summarized as follows:

- The potential evidence is extracted and collected by the digital forensic team. The potential evidence containers are carved, decrypted, de-duplicated and made user-readable using state-of-the-art forensic tools. The resulting data sets are stored in a storage area network (SAN) or using network attached storage (NAS) (note that this can impact data transfer rates). All the servers are managed by the digital forensic team and the original potential evidence containers are secured in a vault.

- The chief of forensic operations accesses the SAN and dispatches data sets according to the case.

- The research team members blind-filter the data and transfer the relevant filtered data to the party counsels.

- The party counsels make the final decisions pertaining to the data (e.g., evidence to be retained, personal data to be discarded and irrelevant data) and store the evidence on a distinct server. The party counsels may also ask the chief of forensic operations to perform additional searches and the research teams to analyze documents using new criteria.

- A virtual private network provides access to the digital forensic team's computer center for situations where several setups are needed for the same e-discovery operation. However, only technical information – not case data – is transmitted via this link.

- At the end of the e-discovery operation, all the hard drives (on laptops, computers, servers) are wiped clean based on DoD standards or are physically destroyed to prevent any data from being recovered.

The e-discovery infrastructure described above is not as "bulky" as it might appear. A lightweight version can be implemented with the servers running locally at the digital forensic team's computer center. Indeed, a mobile version is also feasible.

4.   Conclusions

The modified e-discovery reference model described in this paper is augmented based on our extensive experience with e-discovery cases. The model is simpler than the original EDRM and effectively captures the e-discovery workflow. The information technology architecture based on this model can support the entire e-discovery process and guarantee that non-disclosure agreements and information confidentiality are preserved. A mobile implementation has proved to work very well in real cases.

References

[1] M. Daley and K. Rashbaum (Eds.), The Sedona Conference Framework for Analysis of Cross-Border Discovery Conflicts: A Practical Guide to Navigating the Competing Currents of International Data Privacy and e-Discovery – 2008 Public Comment Version, The Sedona Conference, Sedona, Arizona, 2008.

[2] EDRM, Electronic Discovery Reference Model, St. Paul, Minnesota (www.edrm.net).

[3] R. Losey, e-Discovery Team, Orlando, Florida (www.ralphlosey.wo rdpress.com).

[4] R. McKemmish, What is forensic computing? *Trends and Issues in Crime and Criminal Justice*, no. 118 (www.aic.gov.au/publications /tandi/ti118.pdf), 2002.

VIII

EVIDENCE MANAGEMENT

Chapter 22

CONCEPT MAPPING FOR DIGITAL FORENSIC INVESTIGATIONS

April Tanner and David Dampier

Abstract   Research in digital forensics has yet to focus on modeling case domain information involved in investigations. This paper shows how concept mapping can be used to create an excellent alternative to the popular checklist approach used in digital forensic investigations. Concept mapping offers several benefits, including creating replicable, reusable techniques, simplifying and guiding the investigative process, capturing and reusing specialized forensic knowledge, and supporting training and knowledge management activities. The paper also discusses how concept mapping can be used to integrate case-specific details throughout the investigative process.

Keywords: Concept mapping, investigative process, knowledge management

1.   Introduction

Digital forensic procedures are executed to ensure the integrity of evidence collected at computer crime scenes. Traditionally, the procedures involve the preservation, identification, extraction, documentation and interpretation of computer data [9]. However, due to advancements in technology, digital forensic investigations have moved beyond computers and networks to also encompass portable electronic device, media, software and database forensics [3, 12].

A variety of models have been proposed to improve the digital forensic process; some of the more important ones are investigative models, hypothesis models and domain models. Investigative models focus on the activities that should occur during an investigation [1, 7, 12, 13]. Hypothesis models focus on hypotheses that help answer questions or analyze cases [6]. Domain models concentrate on the information used to examine and analyze cases [3, 4, 14].

G. Peterson and S. Shenoi (Eds.): Advances in Digital Forensics V, IFIP AICT 306, pp. 291–300, 2009.
© IFIP International Federation for Information Processing 2009

A common model for the digital forensic investigative process is not yet available. However, a good candidate is the DFRWS investigative process model [12], which was created by a panel of research experts. The DFRWS model defines six phases in a digital forensic investigation: identification, preservation, collection, examination, analysis and presentation. This paper demonstrates how concept mapping can be used to provide an excellent alternative to the checklist approach used in many investigations. In addition, it shows how case-specific details can be integrated with concept maps produced for the six phases of the investigative process.

2.   Related Work

Venter [17] has proposed a process flow framework to assist first responders during the identification and collection phases of digital forensic investigations. The framework provides a flowchart-based approach for seizing evidence and a centralized mechanism for recording information collected at a crime scene.

Bogen [3] has created a case domain model that provides a framework for analyzing case details by filtering forensically-relevant information. Bogen's model is based on established ontology and domain modeling methods; artificial intelligence and software engineering concepts are used to express the model. The model provides mechanisms for focusing on case specific information, reusing knowledge, planning for examinations and documenting findings.

Kramer [8] has utilized concept maps to capture the tacit knowledge of design process experts. His focus is on collecting, understanding and reusing the knowledge of multiple domain experts in design processes that drive initial design decisions. His approach illustrates the effectiveness of concept maps in eliciting and representing expert knowledge.

Concept maps are a graphical model for organizing and representing knowledge by expressing the hierarchical relationships between concepts. Concept maps were first used to track and understand the scientific knowledge gained by children [11]. Since then, researchers and practitioners from various fields have used them as evaluation tools and decision aids, to plan curricula, to capture and archive expert knowledge and to map domain information [8, 11].

Figure 1 presents a sample concept map, which itself conveys the key features of concept maps. Concepts are represented as enclosed boxes; the lines show how concepts are related to each other. A concept map is similar to a hierarchically, structured checklist in that it provides an organized, structured way to address key points. Unlike checklists,

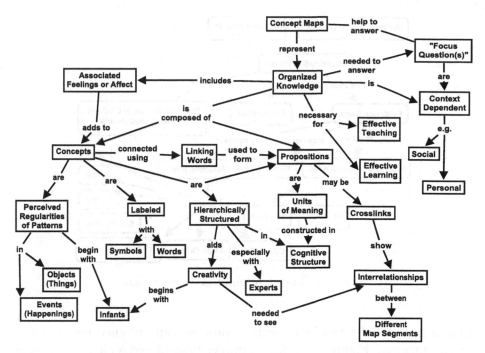

Figure 1. Concept map showing key features of concept maps [11].

however, concept maps show how ideas and concepts are hierarchically linked to each other based on the creator's understanding of the domain. The most inclusive and general concepts are located at the top of the map while more specific concepts are located towards the bottom. Specific event objects, which are not included in boxes, help clarify the meanings of concepts. Prior knowledge of a domain is generally needed to use concept mapping effectively.

Concept maps can be generated manually or using software such as CmapTools. CmapTools, which is used in our work, supports the linking of resources such as photos, images, graphs, videos, charts, tables, texts, web pages, other concept maps and digital media to concepts [11].

Concept maps of the digital forensic investigative process can provide a quick reference of the case domain. Also, they can be used to record case information and to guide novice as well as expert investigators.

3.   Modeling the Investigative Process

The digital forensic investigative process has six phases: identification, collection, preservation, examination, analysis and presentation [12]. Checklists and other documents are commonly used to perform specific tasks associated with each phase. However, applying concept

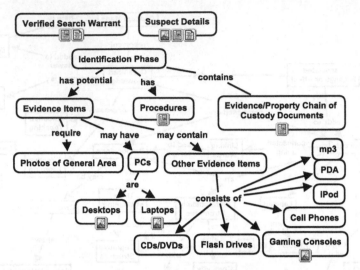

Figure 2. Identification phase concept map.

mapping to model the investigative process can enhance every phase of the process. Figures 2–6, for example, present convenient, graphical views of checklist activities that occur during the key phases of the investigative process. By referring to the concept maps, an investigator can easily determine the actions that should be performed in each phase.

3.1   Identification Phase

The main goal of the identification phase is to determine the items, components and data associated with a crime. The crime scene and evidentiary items should be photographed and documented in detail using proper procedures. According to Kruse and Heiser [9], the computer screen (including open files), the entire computer system and all other potential evidence items should be photographed and documented. Instead of using a checklist for these tasks, Figure 2 provides a graphical view of the steps used to identify digital evidence. The "Chain of Custody Documents" concept shows that specific chain of custody procedures should be followed. The "Evidence Items" concept shows the evidence items that should be identified, and the "Procedures" concept shows the organizational procedures that should be followed.

The concept map can be used by crime scene investigators as a quick reference guide to decide which evidence items should be searched for and as a reminder that the chain of custody should be followed and documented. After all the evidence has been identified and collected, the digital version of the concept map may be augmented to include

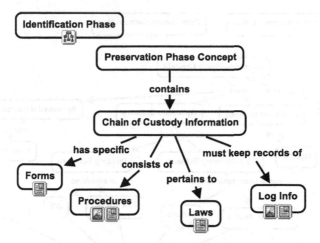

Figure 3. Preservation concept map.

photos, documents and other information obtained from the crime scene. This evidentiary information can be added to the concept map as icons that contain information specific to the case. For example, the "Verified Search Warrant" concept in Figure 2 is associated with an icon that represents a copy of the search warrant. Likewise, the "Photos of General Area" concept could be associated with digital photographs of the crime scene (e.g., computer screen, cabling and network connections).

Note that a concept does not have to be linked to another concept. For example, the "Suspect Details" concept is included in the concept map to provide the investigator with photographs, identifying information and the criminal history of the suspect.

A concept map augmented with icons and related information is very useful for cases that may take years to go to trial. The map could enable an investigator to quickly review the details of the case and the evidentiary items.

3.2   Preservation Phase

Chain of custody is one of the most important tasks associated with the preservation phase [5, 9]. Thorough documentation of the chain of custody helps ensure the authenticity of evidence and refute claims of evidence tampering. It provides complete details about the possession and location of the evidence during the lifetime of a case; these details decrease the likelihood that the evidence will not be admitted in court.

As shown in Figure 3, the chain of custody establishes who collected the evidence ("Forms" concept), how and where the evidence was collected ("Forms" and "Procedures" concepts), who took possession of

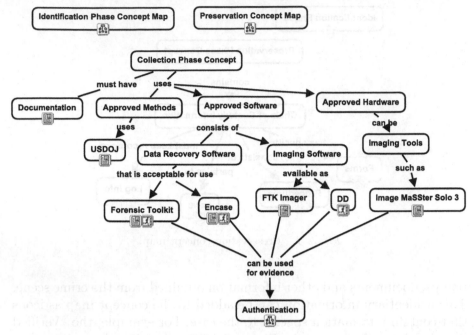

Figure 4.   Collection phase concept map.

the evidence ("Log Info" concept), how the evidence was protected and stored ("Forms" concept), and who removed it from storage and the reasons for its removal ("Log Info" concept) [9]. Other tasks associated with the presentation phase include properly shutting down the computer or evidence item, transporting the evidence to a secure location and limiting access to the original evidence, which are found in the "Procedures," "Forms " and "Log Info" concepts, respectively.

3.3   Collection Phase

Evidence collection involves the use of approved recovery techniques and tools, and the detailed documentation of the collection efforts. All the techniques and tools involved in the evidence collection phase are represented as concepts in Figure 4.

The "Documentation" concept in Figure 4 contains a file icon representing the techniques and tools used to collect the evidence. The tools could be launched from their corresponding concept icons. These icons could also contain links to websites and electronic manuals pertaining to the tools. The identification and preservation phase concept maps (Figures 2 and 3) could be accessed directly from the collection phase concept map as well.

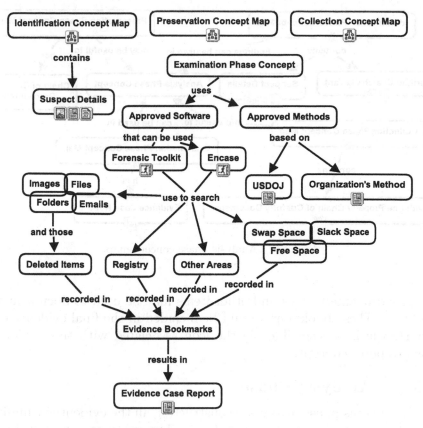

Figure 5. Examination phase concept map.

3.4 Examination Phase

During the examination phase, specialized tools and techniques are used to search for and identify evidence [10]. Evidence may exist in files, emails, images, folders and hidden space on the disk (e.g., slack space, swap space and free space), the registry and other areas as shown in Figure 5. Tools such as the Forensic Toolkit (FTK) and Encase are often used to examine these areas more effectively; these tools also reduce the amount of time spent searching for evidence. Individuals should be trained to operate forensic tools and should use them with utmost care because evidence authenticity is of prime importance.

The "Forensic Toolkit" and "Encase" concepts in Figure 5 have executable icons that could allow the examiner to launch the software and begin examining the evidence. The "Suspect Details" concept is included to accommodate keywords and keyword variations that could assist the examiner in finding more evidence. Bookmarks containing pic-

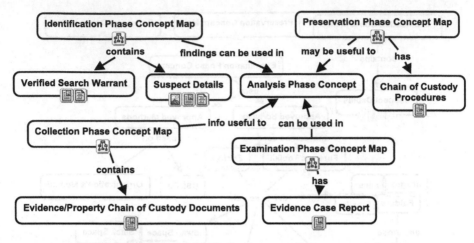

Figure 6. Analysis phase concept map.

tures, video, emails, files and other items may be used to document the evidence. These bookmarks could be included in the final evidence case report, which is accessible via the icon associated with the "Evidence Case Report" concept.

3.5   Analysis Phase

The analysis phase involves reconstructing all the evidentiary findings in order to theorize what occurred [16]. The evidence collected during the examination phase is used to create event timelines, relationships between the evidentiary items and criminal intent.

Concept maps are useful in the analysis phase because they can be used to create event timelines of events from the evidence case report and suspect details. Moreover, concept maps help clarify how the evidentiary items are related to each other. All the evidential findings can be placed in one location and accessed via concept icons. Beebe and Clark [2] state that "data analysis is often the most complex and time-consuming phase in the digital forensic process." Figure 6 provides a good example of how concept maps can provide organization, structure and easy accessibility to the evidence, case details, procedures and chain of custody documentation during the analysis phase.

3.6   Presentation Phase

Every task in the previous five phases plays a role in the presentation of evidence in court. The presentation phase is very important because it is where the legal ramifications of the suspect's actions are determined.

The investigator must be able to show exactly what occurred during the identification, collection, preservation, examination, and analysis phases of the investigation. Often, specialized tools and techniques are used to present the findings in court proceedings [5]. As shown in Figure 6, concept maps provide an attractive alternative for presenting the findings in an organized manner. The investigator would be able to show the court exactly what tasks were performed to obtain the evidence, what evidence was found, and the steps taken to ensure evidence authenticity.

4. Conclusions

Concept mapping can enhance every phase of a digital forensic investigation. The principal benefits are an intuitive graphical view of the investigative process and a simple method for documenting and storing case-specific information such as evidence, case reports, chain of custody documents and procedures. Concept maps also provide a framework for creating a digital forensic repository where case-specific concept maps and specialized techniques can be accessed and shared by the law enforcement community. Other benefits include the ability to uncover misunderstandings in the investigative process, create knowledge management strategies specific to criminal investigations, and provide training and support to novice and expert investigators.

References

[1] V. Baryamureeba and F. Tushabe, The enhanced digital investigation process model, *Proceedings of the Fourth Digital Forensic Research Workshop*, 2004.

[2] N. Beebe and J. Clark, A hierarchical, objectives-based framework for the digital investigation process, *Proceedings of the Fourth Digital Forensic Research Workshop*, 2004.

[3] A. Bogen, Selecting Keyword Search Terms in Computer Forensic Examinations using Domain Analysis and Modeling, Ph.D. Dissertation, Department of Computer Science and Engineering, Mississippi State University, Mississippi State, Mississippi, 2006.

[4] A. Bogen and D. Dampier, Unifying computer forensics modeling approaches: A software engineering perspective, *Proceedings of the First International Workshop on Systematic Approaches to Digital Forensic Engineering*, pp. 27–39, 2005.

[5] D. Brezinski and T. Killalea, RFC3227: Guideline for Evidence Collection and Archiving, Networking Working Group, Internet Engineering Task Force (www.ietf.org/rfc/rfc3227.txt), 2002.

[6] B. Carrier and E. Spafford, An event-based digital forensic investigation framework, *Proceedings of the Fourth Digital Forensic Research Workshop*, 2004.

[7] S. Ciardhuain, An extended model of cybercrime investigations, *International Journal of Digital Evidence*, vol. 3(1), 2004.

[8] M. Kramer, Using Concept Maps for Knowledge Acquisition in Satellite Design: Translating "Statement of Requirements on Orbit" to "Design Requirements," Ph.D. Dissertation, Graduate School of Computer and Information Sciences, Nova Southeastern University, Fort Lauderdale-Davie, Florida, 2005.

[9] W. Kruse and J. Heiser, *Computer Forensics: Incident Response Essentials*, Addison-Wesley, Boston, Massachusetts, 2001.

[10] M. Noblett, M. Pollitt and L. Presley, Recovering and examining computer forensic evidence, *Forensic Science Communications*, vol. 2(4), 2000.

[11] J. Novak and A. Canas, The Theory Underlying Concept Maps and How to Construct and Use Them, Technical Report IHMC Cmap Tools 2006-01, Florida Institute for Human and Machine Cognition, Pensacola, Florida, 2006.

[12] G. Palmer, A Road Map for Digital Forensic Research, DFRWS Technical Report, DTR-T001-01 Final, Air Force Research Laboratory, Rome, New York, 2001.

[13] M. Pollitt, An ad hoc review of digital forensic models, *Proceedings of the Second International Workshop on Systematic Approaches to Digital Forensic Engineering*, pp. 43–54, 2007.

[14] G. Ruibin, T. Yun and M. Gaertner, Case-relevance information investigation: Binding computer intelligence to the current computer forensic framework, *International Journal of Digital Evidence*, vol. 4(1), 2005.

[15] United States Department of Justice, Searching and Seizing Computers and Obtaining Electronic Evidence in Criminal Investigations, Washington, DC (www.usdoj.gov/criminal/cybercrime/s&s manual2002.pdf), 2002.

[16] J. Vacca, *Computer Forensics: Computer Crime Scene Investigation*, Charles River Media, Boston, Massachusetts, 2005.

[17] J. Venter, Process flow diagrams for training and operations, in *Advances in Digital Forensics II*, M. Olivier and S. Shenoi (Eds.), Springer, Boston, Massachusetts, pp. 331–342, 2006.

Chapter 23

SYSTEM SUPPORT FOR FORENSIC INFERENCE

Ashish Gehani, Florent Kirchner and Natarajan Shankar

Abstract   Digital evidence is playing an increasingly important role in prosecuting crimes. The reasons are manifold: financially lucrative targets are now connected online, systems are so complex that vulnerabilities abound and strong digital identities are being adopted, making audit trails more useful. If the discoveries of forensic analysts are to hold up to scrutiny in court, they must meet the standard for scientific evidence. Software systems are currently developed without consideration of this fact. This paper argues for the development of a formal framework for constructing "digital artifacts" that can serve as proxies for physical evidence; a system so imbued would facilitate sound digital forensic inference. A case study involving a filesystem augmentation that provides transparent support for forensic inference is described.

Keywords: Automated analysis, evidence generation, intuitionistic logic

1.   Introduction

As the population density increases, the competition for resources intensifies and the probability of conflict between individuals rises commensurately. The physical and virtual worlds have markedly different mechanisms for managing the potential friction, each clearly influenced by the context in which it was developed. The security requirements of early computing systems were simple enough that they could be precisely specified and implemented [6]. This allowed criminal actions to be prevented before they occurred. For example, if a principal attempted to make an unauthorized change to a document, the reference monitor interceded and disallowed the action. In contrast, principals have greater leeway to violate rules in the physical world. However, they are retroactively held accountable for their actions. The legal system,

G. Peterson and S. Shenoi (Eds.): Advances in Digital Forensics V, IFIP AICT 306, pp. 301–316, 2009.

thus, indirectly deters crime by punishing its perpetrators after they have acted.

The difference in the two approaches to handling crime is fundamental. To proactively prevent crime, it is necessary to characterize what is illegal *a priori*. Security agents must then monitor all activity and intervene when prohibited acts are in progress. An important consequence is that behavior can be policed even when users are pseudonymous. In contrast, the reactive approach relies on losses being reported by victims. The legality of an action can be adjudicated *a posteriori*. This allows complex contexts, such as intent, to be incorporated into the decision about whether specific behavior should be deemed illegal. As the population grows, the resources needed to proactively track every action of every individual become prohibitive. Reactive enforcement requires significantly less effort because the burden of monitoring is distributed among the members of the population (who are potential victims).

As computer systems grow in complexity, the limitations of the proactive model become increasingly apparent. Specifying enterprise-wide security policy is so challenging that corporations routinely outsource the task to specialized consultants [5, 13]. Characterizing attack mechanisms is an unending process, as evidenced by the need for continuous updates of intrusion detection signatures, virus bulletins and stateful firewall rules. Simultaneously, a major impediment to adopting a reactive security model is being overcome. This is the ubiquitous availability of a strong, nonrepudiable notion of identity without which accountability is meaningless. With the help of trusted platform modules [15] in commodity hardware and the mandatory use of cryptographic identities in the next-generation Internet infrastructure [4], global, certified identities will facilitate a transformation in cyber security.

In this paradigm, the creation, discovery and forensic analysis of digital evidence will play a critical role, and we must ensure that the integrity of digital evidence cannot be challenged. Consequently, it is important to create mechanisms that augment digital evidence with artifacts that are difficult to alter without detectable change. By building system support that transparently generates such metadata, digital evidence can be imbued with a level of reliability that exceeds that of evidence gathered from the physical world.

Generating forensically-sound digital evidence allows the use of a reactive security model that would yield three immediate results. First, security policy creators are relieved of the burden of defining exactly who should be allowed to do what in every possible scenario. Instead, they can define policy at a higher level of abstraction. Second, individuals have the freedom to perform a much larger range of actions in the virtual world. As in the physical world, they must be ready to justify their actions if challenged. Third, the legal semantics of particular sequences of actions can be characterized *ex post facto*. This addresses a fundamental weakness in current anomaly detection systems, which incorrectly flag activity as suspicious, especially when they are tuned to be sensitive enough to detect most real crimes. In the reactive paradigm, since no actual crime occurs, no victim will report it; this eliminates the false positives.

Once a loss is reported, the offense must be characterized. Forensic analysis in the physical world relies on the trail of environmental changes made by a crime's perpetrator. In the digital world, criminals could conceivably erase all traces of their activity. It is, therefore, incumbent upon future computing systems to provide trustworthy audit trails with sufficient detail to allow forensic conclusions to be drawn. However, the indiscriminate addition of auditing to a runtime environment introduces performance penalties for executing applications, requires large amounts of storage, and may even compromise privacy. Consequently, it is necessary to determine how to balance the needs of forensic analysis and the users of a system.

2.   Standards of Evidence

The Frye legal standard [3] derives from a 1923 case where systolic blood pressure measurements were used to ascertain deception by an individual. The method was disallowed because it was not widely accepted by scientists. The Frye standard was superseded by the Federal Rules of Evidence in 1975, which require evidence to be based on scientific knowledge and to "assist the trier of fact." In 1993, the Daubert family sued Merrell Dow, claiming that its anti-nausea drug, Bendectin, caused birth defects. The case reached the U.S. Supreme Court, where the Daubert standard [19] for relevancy and reliability of scientific evidence was articulated. In order for digital evidence to be presented in court, the process used to collect it must meet the Daubert standard. In some states, the reliability of the evidence itself must be demonstrable. For example, the Texas Supreme Court extended the Daubert standard in the Havner case [14], ruling that if the "foundational data underlying opinion testimony are unreliable," then they would be considered "no evidence".

Current software systems produce *ad hoc* digital artifacts such as data file headers and audit logs. These artifacts are created by applications that may not meet the standards mentioned above, which reduces their value as evidence in legal proceedings.

We argue that software systems should be developed to automatically emit digital artifacts that cannot be forged, producing fragments whose veracity can be checked during an investigation in the same way forensic analysts currently verify physical evidence found by crime scene investigators. In addition, we suggest that a formal framework should be utilized for analyzing a collection of such digital artifacts since this will effectively codify the set of inferences that can be drawn in a court of law. The combination will allow conclusions from digital forensics to have the same weight as those drawn from physical evidence, once the reliability of digital evidence is established by precedent in court.

3. Challenges

A number of challenges must be addressed in the process of developing a framework for digital forensic analysis.

3.1 Evidence Selection

The first problem is to determine which parts of current proactive protection mechanisms can be transformed into elements in a reactive, accountability-based security apparatus. It is instructive to examine how institutions in the physical world address this issue. When the potential loss is high or the consequence is both likely and irreversible, preventive protection is often utilized. For example, a bank does not leave its vault unguarded and high-ranking public officials in the United States are provided with Secret Service protection since they are likely targets of attack.

Extant legislation already provides relevant guidelines. For example, publicly-traded companies need information flow controls to comply with the Sarbanes-Oxley Act [18], healthcare providers need data privacy protection to comply with the Health Insurance Portability and Accountability Act (HIPAA) [16], and financial services firms and educational institutions have to safeguard personal information to comply with the Gramm-Leach-Bliley Act [17].

A security system can be remodeled so that evidence of relevant activity is generated transparently. For example, instead of specifying and implementing access control for the vast majority of the data in a system, accesses and modifications can simply be recorded. The owner of a piece of data can inspect the corresponding audit trail. An owner who discerns any activity that violates the policy can initiate action against the perpetrator.

3.2   Forensic Analysis

Every crime leaves fragments of evidence. It is up to an investigator to piece the fragments together and create a hypothesis of what transpired. In doing so, the investigator must process the evidence and draw conclusions about the likelihood that the hypothesis is correct. For the operations to be considered forensically sound, at the very least they must be reproducible by the opposing counsel's experts. Consequently, a framework for analysis of the evidence must be agreed upon by all parties.

To see why it is important to have a standardized framework for reasoning about evidence, consider the effect of having different rules for operations that can be performed on digital evidence. If two operations are not commutative, then their composed output can be challenged on the grounds of the order in which they were performed. If there is agreement on commutativity, then the operations can be arbitrarily composed and the output would be considered to be acceptable. Similarly, if an operation is accepted as invertible, its input and output can be compared and checked for consistency. Any inconsistency can serve as grounds for having evidence discarded. In contrast, if an operation is not deemed to be invertible and the output is unimpeachable, then the absence of consistency between an input and output would not be grounds for eliminating the input from consideration as evidence.

Operations must be repeatable in order to meet the Daubert standard for scientific evidence. A digital forensic system must be designed to allow efficient distinctions to be made about which evidentiary properties are satisfied. For example, if a piece of evidence was derived using randomness, user input or network data, its legal admissibility will differ from the content that can be completely recomputed when persistent files are used as inputs.

3.3   Chain of Custody

When a piece of evidence is to be presented in a court, the chain of custody of the evidence must be established to guarantee that it has not been tampered with. The process makes two assumptions that do not hold by default in the virtual world. The first is that the evidence was not altered from the time it was created to the time it was collected. In a world where data is rapidly combined to produce new content, it is likely that the data found during an investigation will have already undergone editing operations before it was collected as evidence. The second assumption is that a piece of evidence was created by a single individual. A virtual object is much more likely to have multiple co-authors. Note that a co-author is a principal who owns one of the processes that operated on any of the data used to create the object in question.

In principle, these issues can be addressed by designing software functionality that transparently annotates data with the details of its provenance. If the metadata generated is imbued with nonrepudiable authenticity guarantees, it can serve as forensic evidence to establish a chain of custody. A policy verification engine can be used to infer the set of co-authors of a piece of data by inspecting a set of metadata provided as an input. It can also follow the chain of metadata attestations about modifications of the data to ensure that its evidentiary value is not negatively impacted.

4.   Formal Framework

The utilization of a formal framework with an explicitly defined logic has a number of advantages over *ad hoc* analysis of digital evidence.

4.1   Standardization

The set of laws that govern forensic evidence handling and inference can be codified in the rules of the logic. The variations and precedents of each legal domain, such as a state or county, can be added as a set of augmenting axioms and rules. In particular, such standardization allows the prosecution and the defense to determine before trial what conclusions will likely be drawn by expert witnesses in court. Further, the significance of a digital artifact can be tested by checking which conclusions are dependent upon it. Thus, standardization may decrease the time to arrive at an agreement in court about which conclusions can be drawn given a body of digital evidence.

4.2   Automation

A framework that is completely defined by a formal logic can serve as a technical specification for implementing a forensic inference engine in software. In the physical world, the number of pieces of evidence introduced in court may be limited. In contrast, the number of pieces of digital evidence that may need to be utilized to draw a high-level conclusion may be significantly larger. In either case, as the number of elements that must be assembled to draw a conclusion increases, the effort to construct a sound inference grows exponentially. Automating the process will become necessary to ensure that the cost of using a body of digital evidence remains affordable in cases where the plaintiff or the defendant has a limited budget.

4.3   Soundness

Automating the process of generating nonrepudiable digital artifacts in software is likely to generate large bodies of digital evidence usable in a court of law. If the evidence must be manually assembled into a chain of inference, the likelihood of erroneous conclusions could be significant. All the pieces of evidence must be arranged into a plausible timeline, requiring numerous alternative orderings to be evaluated. Further, certain conclusions can be ruled out because the supporting evidence may be contradictory, such as being predicated on a person being in two locations at one time.

Manually verifying complex properties is likely to introduce errors that may be too subtle to identify without investing substantial resources. Automating the forensic inference process with an explicit formally defined framework guards against the introduction of such errors and ensures that the conclusions are sound.

4.4   Completeness

A formal framework that is complete yields a set of theorems that are the only possible conclusions logically inferred from the set of axioms determined by the digital evidence. If an attempt is made to draw any other conclusion, a judge can use the completeness of the forensic inference system to justify setting aside an argument on the grounds that it does not follow from the evidence.

Given a set of elements corresponding to the digital evidence and a decidable logic, an automated theorem prover can generate a sequence of all possible theorems, each corresponding to a conclusion for which a proof is available. A lawyer can examine the theorems (after filtering using suitable constraints if there are too many to inspect) to see if any of them either corroborate a hypothesis or to search for new hypotheses not previously considered. Having exhausted the set of theorems produced, the lawyer will be assured of not missing any possible line of argument using the available evidence.

5.   CyberTrail

Our framework for generating and reasoning about digital artifacts is called CyberTrail. It utilizes CyberLogic [2] to reason about digital evidence. CyberLogic allows protocols to be specified as distributed logic programs that use predicates and certificates to answer trust management queries from the available evidence. Since proofs of claims are constructive, every conclusion is accompanied by an explicit chain of evidence. The logic is general enough to be utilized in a broad set of applications. In particular, its properties make it well suited for use by CyberTrail, as described below.

5.1   Digital Artifacts

A key aspect of CyberTrail is its proactive generation of digital artifacts that are as reliable as evidence gathered from the physical world. While there is no assurance that an artifact corresponding to any particular event of interest will be generated, the artifacts produced must be difficult to forge. Otherwise, the creator of an artifact could repudiate it on the grounds that someone else may have invested the effort to falsely generate the artifact.

We utilize the CyberLogic primitive for attestation to generate non-repudiable digital artifacts. Every authority that generates attestations must have an associated signing key and verification key pair. An authority can be any entity, from a user to a piece of software. An attestation denoted by $A \!:\!\triangleright\! S$ indicates that statement S has been signed cryptographically by authority A (using a digital signature algorithm and the signing key of authority A).

By generating digital artifacts that attest to the state and operations of a small subset of a system, CyberTrail seeks to leave a nonrepudiable trail that would allow the forensic analyst to make inferences about the global state and operations of the system. This is analogous to gathering physical evidence from a crime scene to reconstruct the events that led up to the activity that transpired during the actual crime.

5.2   Constrained Claims

Since software systems can make arbitrary claims by emitting a predicate, it is necessary to ensure that the authority to make a claim is captured in CyberTrail. This is accomplished by leveraging CyberLogic's support for bounded delegation. $\mathcal{D}_n(A_{from}, A_{to}, S)$ is generated by an agent A_{from} to indicate that A_{to} is authorized to make statement S on A_{from}'s behalf. The subscript n denotes how many successive levels A_{to} is allowed to delegate the right. By making authorization explicit, statements produced by software without accompanying evidence of delegation are constrained.

5.3   First-Order Logic

Propositional logic is not expressive enough to capture relationships where the variables must be quantified. For example, checking whether any file in the evidence was owned by a specific user u is easily expressed in first order logic as $\exists f\ User(u) \wedge File(f) \wedge Owner(u, f)$ where the predicate $File(f)$ is true if f is a file; the predicate $User(u)$ is true if u is a user; and the predicate $Owner(u, f)$ is true if user u owns file f. CyberLogic can be extended to use higher-order logic if necessary.

5.4   Intuitionistic Logic

Classical logic, dating back to Aristotle, includes the Law of Excluded Middle, which says that a statement is either true or false (and that there is no third possibility); i.e., $S \vee \neg S$ is a tautology, where S is a statement in the logic. The original rule pertained to statements about finite sets and its was subsequently extended to infinite sets [20]. However, this gave rise to contradictions like Quine's Liar Paradox [10]. Intuitionistic logic [8] removes the Law of Excluded Middle from classical logic. The result is that the logic is better able to model the ambiguity of the real world.

Consider the statement that every user owns a file, which can be formulated as $\forall u\ \exists f\ User(u) \wedge File(f) \wedge Owner(u, f)$. Given a particular user u and a fixed set of files, it is possible to determine whether $Owner(u, f)$ holds for each f in the set. However, in general, the set of all files is not known, so it is not possible to determine whether $\exists f\ User(u) \wedge File(f) \wedge Owner(u, f)$ holds or whether $\neg(\exists f\ User(u) \wedge File(f) \wedge Owner(u, f))$ holds. If the statement $S = \exists f\ User(u) \wedge File(f) \wedge Owner(u, f)$, then in classical logic, $S \vee \neg S$ would need to hold. This property does not have to hold in intuitionistic logic, which allows us to reason about situations where the universe is open. This is important when dealing with evidence because artifacts are not limited to originating from a predefined closed universe.

5.5   Temporal Modality

A statement may hold in a limited context rather than being universally true. An authority who makes a claim about a statement may wish to convey the context explicitly. Modal logic introduces the $\square$ and $\diamond$ operators to indicate whether a statement is "necessarily" or "possibly" true, respectively. Temporal logic allows the validity of the statement in time to be specified. The utility is apparent when considering how to qualify the validity period of a digital certificate. CyberLogic allows an attestation to take the form $A :\!\triangleright_{=t} S$ to indicate that it holds at time t. This suffices for constructing other modalities of attestation by quantifying the time. For example, $\square A :\!\triangleright S$ becomes $\forall t\ A :\!\triangleright_t S$.

6.   Case Study

This section describes a case study involving a filesystem augmented with CyberTrail features.

6.1   User-Space Filesystem

FUSE[12] provides a Linux kernel module that intercedes when filesystem calls are made. The call and its arguments are passed to a user-space daemon by communicating through a special character device defined for the purpose. The call can then be handled by a user-defined function. Our current prototype augments a subset of the VFS filesystem API and constructs the metadata needed for CyberTrail predicates. However, in its current form, information is inserted as records into a relational database using SQL commands. Reasoning about this digital evidence would require custom query tools built atop the SQL query interface.

Developing a new filesystem in user-space allows end users to utilize it without having to modify the current filesystem. Legacy applications can be executed and are presented with the same interface while the user-space filesystem acts as a transparent layer between the application and the native filesystem. Furthermore, errors in implementation do not crash the kernel or corrupt portions of the native filesystem that were not directly operated upon.

Since the prototype operates in user-space, it can only generate facts signed by the user (which are useful to ensure that the user does not subsequently make claims that could be repudiated by that user's earlier claims). A future extension could leverage a trusted platform module [15] to transparently construct facts without the cooperation of the user (to address a broader range of threats).

6.2   Transparent Reasoning

The next step in developing CyberTrail in the context of a filesystem is to marry the predicate generation and attestation directly with an automated reasoning environment. Every CyberLogic statement is a hereditary Harrop formula, the logical construct that is the basis for λ-Prolog [9]. We could invoke the λ-Prolog interpreter when the FUSE user-space daemon starts, and then insert predicates and attestations when the modified filesystem calls execute. By defining an inter-node communication protocol between distributed instances of the λ-Prolog interpreter, queries could be automatically resolved even when data has been modified at multiple nodes and transferred between them. In particular, subgoals could be resolved at the nodes corresponding to their targets, transparently guiding the distribution of the resolution procedure.

6.3   Granularity

If all the input data, application code and system libraries used to generate an output are available, an operation can be verified by repeating it. In practice, programs often read data from ephemeral sources such as network sockets, filesystem pipes and devices that provide sources of randomness. This prevents the output of the program from being independently validated in a distributed environment because the verifier must trust the original executor's claims about the contents derived from these sources. Since the executor has the freedom to alter the ephemeral inputs to yield a preferred output, checking the operation by repeating it does not increase the likelihood that the claimed operation was the one that was previously performed. Hence, our checks are restricted to the persistent files that are read and written by a process.

6.4   Auditing

When the system boots, an audit daemon is initialized. This maintains a table that maps process identifiers to associated metadata. The entry corresponding to each process entry contains an `accessed` list of all files that have been read by it, and a `modified` list of all files that have been written by it.

It is necessary to intercede on file `open()`, `close()`, `read()` and `write()` system calls. When a file `open()` call occurs, a check is performed to see if the calling process has an entry in the table. If not, an entry is created and populated with empty `accessed` and `modified` lists. When a `read()` operation occurs, the file being read is added to the `accessed` list of the calling process. Similarly, when a `write()` occurs, the file is added to the `modified` list of the calling process.

When a `close()` occurs, the `modified` list of the calling process is checked. If the file has actually been written to (as opposed to just being opened for writing or just having been read), the `modified` list will contain it. In this case, the `accessed` list of the process is retrieved from the table. It contains the list of files $\{i_1, \ldots, i_n\}$ that have been read during the execution of the process up to the point that the output file o was closed.

6.5   Artifact Generation

CyberTrail generates a variety of logical facts. We define a filesystem "primitive operation" to be an output file, the process that generated it and the set of input files it read in the course of its execution. For example, if a program reads a number of data sets from disk, computes a result and records it to a file, a primitive operation has been performed.

A primitive operation can be described as follows. Let o be the output file of the operation executed by user e using input files $i_1, \ldots, i_n$. If a process writes to a number of files, a separate instance of a primitive operation would represent each output file. Assume the predicates of Section 5.3 and that $Process(p)$ is true if p is a process; $Output(p, o)$ is true if file o has been written by process p; $Input(p, i)$ is true if file i has been read by process p; and $Owner(e, p)$ is true if process p has been executed by user e.

In practice, a file identifier i has the form $i = (h, f, t)$ where h is the hostname on which the file with name f was last modified at time t, in order to disambiguate files on different nodes with the same name as well as to differentiate between the state of the contents of a file at different instants of time. The identity e must have global meaning. We assume the availability of a public key infrastructure [7]. However, any distributed mechanism for resolving identities, such as linked local namespaces [1] or a web of trust [11], can be used instead.

The facts that correspond to the primitive operation are listed below. Note that CyberTrail must emit them as facts so that they can be used to resolve queries.

$$
\begin{array}{c}
Process(p) \\
Owner(e, p) \\
File(o) \\
Output(p, o) \\
File(i_1) \\
Input(p, i_1) \\
\vdots \\
File(i_n) \\
Input(p, i_n)
\end{array}
$$

The above step would occur after all references to the file become inactive. This possibility arises since multiple valid concurrent references may result after a single open() call. Such a situation occurs when a process spawns multiple threads and passes them a file descriptor. Equivalently, this occurs when a process makes a fork() call, creating another process that has copies of all its active file descriptors. Alternatively, this can occur if a part of the file was mapped to memory. Once all the active file descriptors are closed and the relevant memory blocks are unmapped, digital artifact generation can proceed to completion.

Digital artifacts must be difficult to forge. Therefore, the set of facts emitted above cannot serve as artifacts. They can be used during the process of forensic analysis to determine what activity had occurred in the system, but a suitable set of artifacts must also be found and assembled to validate the reconstruction. The set of digital artifacts that must be generated to accompany the set of facts listed above is given below.

$$e :\triangleright Owner(e, p)$$
$$e :\triangleright Output(p, o)$$
$$e :\triangleright Input(p, i_1)$$
$$\vdots$$
$$e :\triangleright Input(p, i_n)$$

6.6   Forensic Analysis

If CyberTrail functionality is deployed ubiquitously in software systems, the likelihood of finding digital artifacts generated by the operating system or subsequently by applications as well will increase. When a crime occurs and computer systems are involved, a forensic analyst will be able to scour the systems for digital artifacts and enter them into a database of evidence.

A forensic analyst may issue a variety of queries to the database of evidence. For example, the analyst may wish to determine the chain of custody for a piece of data. We use a query language with Prolog semantics to demonstrate the query. If i_0 denotes the file at beginning of the period of interest and i_1 denotes the same file at the end of the period of interest, the following query would verify that a complete chain of custody is available in the database:

$$Chain(i_0, i_1) :=$$
$$Chain(i, i_1) \;\wedge\; Output(p, i) \;\wedge\; Input(p, i_0) \;\wedge$$
$$e :\triangleright Output(p, i) \;\wedge\; e :\triangleright Input(p, i_0)$$

On the other hand, if the analyst wishes to check if all the users who modified a particular file are known, the following query could be issued:

$$Authors(i_0) :=$$
$$Output(p, i_0) \wedge Owner(e, p) \wedge Input(p, i_1) \wedge Authors(i_1)$$

Note that the above query does not validate the query against digital artifacts. If this is required, the query would be extended to:

$$Authors(i_0) :=$$
$$Output(p, i_0) \wedge Owner(e, p) \wedge Input(p, i_1) \wedge Authors(i_1) \wedge$$
$$e \rhd Output(p, i_0) \wedge e \rhd Owner(e, p) \wedge e \rhd Input(p, i_1)$$

CyberTrail also enables a forensic analyst to find all the files that were derived from a particular piece of data. The following query would be issued:

$$Derivatives(i_0) :=$$
$$Input(p, i_0) \wedge Output(p, i_1) \wedge Derivatives(i_1)$$

Similarly, a query could be constructed to check if a particular user had modified any of the data incorporated into a file. By adding more facts and artifacts, such as details of the runtime environment of a process at the time of auditing, other types of queries may be formulated. For example, a forensic analyst may wish to find all the files that were modified by an email client running under a suspect's identity during a given time period. Such a query could be constructed if the facts for the $Command(p)$ predicate were to be introduced at audit time, where $Command(p)$ is the command line used to create the process p.

7. Conclusions

Digital evidence is becoming increasingly important, but is often not sound enough to withstand court challenges. The approach described in this paper produces nonrepudiable digital artifacts in the context of filesystem operations that would enable an investigator to answer a number of forensic queries. Extensions to other software systems are complementary, enabling further inferences to be drawn from the resulting digital artifacts.

Acknowledgements

This research was partially supported by the National Science Foundation under Grant Nos. OCI-0722068 and CNS-0644783.

References

[1] M. Abadi, On SDSI's linked local name spaces, *Journal of Computer Security*, vol. 6(1-2), pp. 3–21, 1998.

[2] V. Bernat, H. Ruess and N. Shankar, First-Order CyberLogic, Technical Report, SRI International, Menlo Park, California (ftp.csl.sri .com/pub/users/shankar/cyberlogic-report.pdf), 2005.

[3] Court of Appeals of the District of Columbia, Frye v. United States, *Federal Reporter*, vol. 293, pp. 1013–1014, 1924.

[4] GENI Project Office, Global Environment for Network Innovations, BBN Technologies, Cambridge, Massachusetts (www.geni.net).

[5] International Business Machines, Security policy definition, Armonk, New York (www-935.ibm.com/services/us/index.wss/offer ing/gbs/a1002391).

[6] B. Lampson, Protection, *ACM Operating Systems Reviews*, vol. 8(1), pp. 18–24, 1974.

[7] U. Maurer, Modeling a public key infrastructure, *Proceedings of the Fourth European Symposium on Research in Computer Security*, pp. 325–350, 1996.

[8] J. Moschovakis, Intuitionistic logic, *Stanford Encyclopedia of Philosophy*, Metaphysics Research Laboratory, Stanford University, Palo Alto, California (plato.stanford.edu/entries/logic-intuitionistic).

[9] G. Nadathur, A proof procedure for the logic of hereditary Harrop formulas, *Journal of Automated Reasoning*, vol. 11(1), pp. 115–145, 1993.

[10] W. Quine, *The Ways of Paradox*, Harvard University Press, Cambridge, Massachusetts, 1962.

[11] M. Reiter and S. Stubblebine, Toward acceptable metrics of authentication, *Proceedings of the IEEE Symposium on Security and Privacy*, pp. 10–20, 1997.

[12] SourceForge, FUSE: Filesystem in userspace (fuse.sourceforge.net).

[13] Sun Microsystems, Security policy services, Santa Clara, California (www.sun.com/service/security/securitypolicyservices.xml).

[14] Supreme Court of Texas, Merrell Dow Pharmaceuticals, Inc. v. Havner, *South Western Reporter*, vol. 953(S.W.2d), pp. 706–733, 1998.

[15] Trusted Computing Group, Beaverton, Oregon (www.trustedcomp utinggroup.org).

[16] U.S. Government, Health Insurance Portability and Accountability Act, Public Law 104–191, *United States Statutes at Large*, vol. 110(3), pp. 1936–2103, 1997.

[17] U.S. Government, Gramm-Leach-Bliley Act, Public Law 106–102, 106th Congress, *United States Statutes at Large*, vol. 113(2), pp. 1338–1481, 2000.

[18] U.S. Government, Sarbanes-Oxley Act, Public Law 107–204, 107th Congress, *United States Statutes at Large*, vol. 116(1), pp. 745–810, 2003.

[19] U.S. Supreme Court, Daubert v. Merrell Dow Pharmaceuticals, Inc., *United States Reports*, vol. 509, pp. 579–601, 1983.

[20] J. van Heijenoort, *From Frege to Godel: A Source Book in Mathematical Logic 1879-1931*, Harvard University Press, Cambridge, Massachusetts, 1967.